ROUTLEDGE LIBRARY EDITIONS:
COMPARATIVE URBANIZATION

Volume 10

# SUBURBIA

# SUBURBIA

## An International Assessment

DONALD N. ROTHBLATT
AND DANIEL J. GARR

LONDON AND NEW YORK

First published in 1986 by Croom Helm Ltd

This edition first published in 2021
by Routledge
2 Park Square, Milton Park, Abingdon, Oxon OX14 4RN

and by Routledge
52 Vanderbilt Avenue, New York, NY 10017

*Routledge is an imprint of the Taylor & Francis Group, an informa business*



*British Library Cataloguing in Publication Data*
A catalogue record for this book is available from the British Library

ISBN: 978-0-367-75717-5 (Set)
ISBN: 978-1-00-317423-3 (Set) (ebk)
ISBN: 978-0-367-77243-7 (Volume 10) (hbk)
ISBN: 978-0-367-77256-7 (Volume 10) (pbk)
ISBN: 978-1-00-317046-4 (Volume 10) (ebk)

**Publisher's Note**
The publisher has gone to great lengths to ensure the quality of this reprint but points out that some imperfections in the original copies may be apparent.

**Disclaimer**
The publisher has made every effort to trace copyright holders and would welcome correspondence from those they have been unable to trace.

# SUBURBIA
## AN INTERNATIONAL ASSESSMENT

Donald N. Rothblatt
and Daniel J. Garr

CROOM HELM
London & Sydney

Croom Helm Ltd, Provident House, Burrell Row,
Beckenham, Kent BR3 1AT
Croom Helm Australia Pty Ltd, Suite 4, 6th Floor,
64-76 Kippax Street, Surry Hills, NSW 2010, Australia

British Library Cataloguing in Publication Data

Rothblatt, Donald N.
Suburbia: an international assessment.
1. Suburbs
I. Title II. Garr, Daniel J.
307.7'4 HT351

ISBN 0-7099-2258-2

Printed and bound in Great Britain
by Billing & Sons Limited, Worcester.

CONTENTS

Contents

## TABLES

FIGURES

# ACKNOWLEDGMENTS

Many individuals, public institutions and representatives of private business helped us make this book possible. In gratitude, we would like to thank Deana Dorman, Mara Southern, Jo Sprague and Serena Wade for their advice and cricitism of the original conceptualization and design of the San Jose case study; Rachelle Alterman, Moshe Braver, Thelma Duchan, Elisha Efrat, Akiva Flexer, Audrey Gil, Ahron Hibshoosh, Michael Romann, Arie Shacher and Schmuel Yavin for their advice and assistance with the Israeli study; Sandra Boes, F. D. J. Bootsma, J. Den Draak, Mart Tacken and Jan Van Weesep for their help with the Dutch study. Our special thanks are also due to Timothy Chalmers, Zelko Pavic, Karen Schirle, and Patricia Stadel for their assistance with the data analysis of our study. In addition, we would like to thank the many graduate students at San Jose State University, Tel-Aviv University and The Hebrew University of Jerusalem, and the firm of Ogilvie Marktonderzoek, B.V., of Amsterdam, who helped us conduct our field work. Last, but not least, we are indebted to the thirteen hundred women on three continents who were willing to be interviewed and share a part of their suburban world with us.

We are also grateful to our colleagues in the Departments of Urban and Regional Planning and Communications Studies at San Jose State University, especially Jo Sprague; to the former Coordinators of the Women's Studies Program, Ellen Boneparth and Sybil Weir; and to the past and present Deans of the School of Social Sciences, Gerald Wheeler and Charles Burdick, for their support and encouragement throughout this study.

In addition we are grateful to the California State University and Colleges which provided funds for Research Assistants; the National Science Foundation which provided financial assistance with Faculty Research Grants in the United States; Tel-Aviv University and the Hebrew University of Jerusalem for facilitating our Israeli study with a Lady Davis Visiting Professorship; the Institute for Town Planning Research of the Delft University of Technology, for supporting our research in the Netherlands, and the Council for the International Exchange of Scholars and the Netherlands-American Commission for their generous funding support in the form of a Senior Fulbright Research Award.

We would like to express our appreciation to Christine Haw, Ann Rothblatt and Syreeta Lachat for their invaluable editorial assistance. Special thanks also go to Estelle Akamine, Margaret Asuncion, and Annelise Bazar for their diligent typing and retyping of the original manuscript. We, of course, accept responsibility for all errors.

For Our Parents

Sophie, Harry, Flora, Florence, and Morris

# INTRODUCTION

Few planning topics have proven more intriguing and controversial than the ongoing post-1945 suburban boom, which is now only beginning to wane after four decades of vigor. Yet, there are few serious studies of the impact of suburbia on its residents and consequently on society as a whole despite the fact that suburban metropolitan development is now a significant form of urbanized living in industrial societies.[1] Nevertheless, high interest rates, energy and transportation costs, and changes in family composition and in the status of women have compelled many suburbanites and would-be suburbanites to re-evaluate the benefits and costs of their living situations.

This book will provide a framework for making such a re-evaluation on an international scale by examining the opportunities for creating satisfying living environments in the suburbs of metropolitan areas in three industrialized countries: the United States, Israel, and the Netherlands. First, the study will present an overview of metropolitan decentralization and suburban development in the industrialized world since 1945. The second chapter of the book will examine the social, economic, political and physical aspects of suburbanization in these three societies which are representative of suburban patterns in the industrialized world. This chapter will focus on the recent changing forces influencing suburbanization and their implications for metropolitan development.

The third chapter will present the methodology for a cross-national study which measures levels of satisfaction with various dimensions of suburban life for families residing in a range of social and physical environments characteristic of much of suburban life in the United States, Israel and the

Netherlands. The dimensions to be examined will be variables concerned with satisfaction with four of the major aspects of the quality of metropolitan life: housing and the physical environment; community services; social patterns; and psychological well-being. We will also attempt to determine and measure population and environmental characteristics (independent variables) such as education, income, housing density and design features, which seem to be related to such levels of satisfaction (dependent variables). That is, for each country, we intend to test the relationships between variables which have been derived from theories and previous research in order to explain the variations anticipated in levels of satisfaction with suburban life. In addition, we will make a comparative analysis of the findings of our study in the three countries.

Questions concerning each variable were designed to yield an ordinal satisfaction score and put into the form of a questionnaire schedule. Utilizing this survey instrument, women were interviewed from households typical of metropolitan populations in each of the three nations. In all, seventeen neighborhoods were surveyed: eight in the San Jose metropolitan area; five in the Tel-Aviv and Jerusalem regions; and four in The Hague agglomeration. Partial correlation analysis and other techniques will be used to test expected relationships between the independent and dependent variables.

Findings and discussion of the study will be presented in the fourth chapter. The discussion will address the major questions underlying our study: for each country, what is the relative importance of environmental influences (e.g., neighborhood design features, location) and social influences (e.g., social class, subcultural, life cycle) on the satisfaction that households have with the suburban life.

Chapter five will summarize the cross-national generalizations made from our comparative findings and examine the implications of our case study and literature review for urbanization in industrialized countries. Here we will focus on the public policy implications of our research for improving the quality of metropolitan living environments in the industrialized world.

The book will close with a discussion of the implications of our work for cross-national comparative urban research. In this last chapter we will stress the implications of our project for future cross-national research as a means of deepening our understanding of metropolitan development in various social, economic, and institutional settings in the industrialized world.

## NOTES

[1] Some of these recent works are: Claude S. Fischer, To Dwell Among Friends: Personal Networks in Town and City (The University of Chicago Press, Chicago, 1982); Aviva Lev-Ari, 'Spatial Behavior and Overt Residential Preference', unpublished Master's Thesis, Hebrew University of Jerusalem, Jerusalem, 1978; William Michelson, Environmental Choice, Human Behavior and Residential Satisfaction (Oxford University Press, New York, 1977); David Popenoe, The Suburban Environment: Sweden and the United States (The University of Chicago Press, Chicago, 1977); Donald N. Rothblatt, Daniel J. Garr and Jo Sprague, The Suburban Environment and Women (Praeger, New York, 1979); and Robert B. Zehner, Indicators of the Quality of Life in New Communities (Ballinger, Cambridge, Mass., 1977).

# CHAPTER 1

# SUBURBIA IN THE INDUSTRIALIZED WORLD: METROPOLITAN DECENTRALIZATION SINCE 1945

Throughout human evolution we have developed an adaptive huddling quality, a need to be near others, at first tentative and cyclical, and then gradually permanent in character. From the first cave dwellings and simple hamlets, to the splendor of capitals of vast empires, to the rapid dynamics of industrial centers, close physical proximity was essential for not only collective survival and social and spiritual needs, but for the efficient production, distribution and consumption of the ever changing goods and services desired by an expanding urban society.[1]

Yet, Lewis Mumford points out that patterns of population dispersion in the form of 'the suburb becomes visible almost as early as the city itself'.[2] Indeed, evidences of suburban development have even been discovered outside the walls of ancient Mesopotamian cities, such as Ur.[3] But suburban environments - essentially outlying residential areas adjacent to and usually dependent on the central city - were available primarily for the very privileged throughout most of urban history.[4]

## A. EARLY SUBURBAN DEVELOPMENT IN THE INDUSTRIALIZED WORLD

It was not until the second half of the 19th century that suburbia even became a model of the good life for the average household as an alternative to the evils of the industrial city. This phenomenon was made possible by transportation innovations, such as the establishment of commuter trains and streetcar lines, which opened many new sites for development beyond the city limits. An economic imperative was supplied by the emerging middle classes of industrial countries with their desires to find their own healthful and peaceful environments in the hinterland.[5] Indeed, suburban and new town conceptualizations became the cornerstone of Anglo-American planning ideology - an ideology still with us today in much of the industrialized world, and one which has generated the image of a pastoral garden city with safe residential environments suitable for families with children.[6]

The outward movement of the central city population to the suburbs in the most industrialized countries was further reinforced with the advent of automobile usage during the 1920s and 1930s which opened up still larger areas for suburban development for middle income families. By the end of the 1930s, metropolitan areas - central cities tied to a ring of expanding suburban communities - were an established and growing phenomenon in advanced nations. By 1940 the United States had become primarily metropolitan, with 51.1 percent (67 million) of its national population living in 147 metropolitan areas[7] while England had about 40 percent of its national population in its six major conurbations.[8] As Lewis Mumford observed:

> By the middle of the twentieth century there were a host of new metropolitan areas, with bulging and sprawling suburban rings that brought many more (people) within the general metropolitan picture.[9]

B. SUBURBAN GROWTH SINCE 1945

While suburbanization did develop substantially around the major cities of industrial countries during the late 19th and early 20th centuries, its most vigorous expression awaited the end of World War II when economic, technological and social forces enabled suburbia to become a major, if not predominant mode of middle class life. In most industrialized countries the pent-up demand for housing during the depression and war years, the baby boom, government programs that encouraged suburban home-building (such as subsidies for mortgages or rents and major highway or rail projects), the desire of the younger middle class to escape the central city, and the mutually reinforcing factors of rising real incomes, industrial decentralization, and the widespread use of the automobile all inexorably accelerated the outward push of metropolitan development on an unprecedented scale.[10]

This process began almost immediately after World War II in those affluent nations least affected by the large scale destruction and social and economic dislocations of the war. In the United States, for example, spectacular suburban population booms occurred in many outlying residential developments such as the Levittowns during the late 1940s and 1950s, when the growth rate outside of the central city exceeded that within by a factor of almost five (see Table 1.1).[11] Thus, while nations like the United States and Canada were well into the process of widespread suburbanization, European countries and Japan were still primarily engaged in the rebuilding of their economies and central cities.[12] As Peter Self pointed out in his comparative study of regional planning:

> After 1945 many European cities embarked upon ambitious schemes of redevelopment and expansion. This was an era of big city development machines backed by highly professional staffs and political leaders.[13]

However, as massive programs to rebuild central cities unfolded, it became increasingly difficult for these cities to accommodate the growing number of urban residents generated by the high rates of natural increase in many urban areas. In addition, migration from outlying, less developed rural areas, and the unprecedented flow of labor from Mediterranean Europe and North Africa to northern and western Europe further intensified demographic pressure on central cities.[14] As Peter Hall and Dennis Hay commented about European urban patterns during the 1950s:

> . . . population was concentrating remarkably into metropolitan cores: a process of centralization was taking place - above all by movements from the non-metropolitan rural periphery.[15]

As Table 1.2 indicates, the European urban cores captured nearly two-thirds of the net population growth during the 1950-60 period. Similar patterns were also noted in other industrial parts of the world such as Japan, despite a staggering decline in its wartime urban population.[16] Consequently, post-1945 central cities affected by the war were dealing with the problem of limited

TABLE 1.1 CENTRAL CITY AND SUBURBAN POPULATION CHANGES IN METROPOLITAN AREAS IN THE UNITED STATES: 1900-80

| | Percent Change in Metropolitan Population [a] | | |
|---|---|---|---|
| Time Periods | Central Cities | Suburban Rings Outside Central Cities | Total |
| 1900-10 | 35.3 | 27.6 | 32.6 |
| 1910-20 | 26.7 | 22.4 | 25.2 |
| 1920-30 | 23.3 | 34.2 | 27.0 |
| 1930-40 | 5.1 | 13.8 | 8.3 |
| 1940-50 | 13.9 | 34.7 | 21.8 |
| 1950-60 | 10.7 | 48.6 | 26.4 |
| 1960-70 | 6.4 | 26.8 | 16.6 |
| 1970-80 | 0.6 | 17.4 | 9.4 |

Note: a. Standard Metropolitan Statistical Areas (SMA's) as defined at the dates indicated.

Source: 1900-1950 data from Donald J. Bogue, Population Growth in Standard Metropolitan Areas 1900-1950, (Housing and Home Finance Agency, Washington, D.C., 1953)
1960-70 data from Census of Population, (U.S. Department of Commerce, Bureau of the Census, Washington, D.C., 1960, 1970)
1980 data from Preliminary 1980 Census, (U.S. Department of Commerce, Bureau of the Census, Washington, D.C., March, 1981)

land resources in the face of overwhelming population pressure. At the same time, redevelopment of these central cities often reduced their residential carrying capacity because, as Peter Self points out, 'overcrowding was eliminated, new roads provided and new standards applied to space around buildings and to schools, playing fields and so forth'.[17]

In addition, other land use shifts were occurring in many city centers such as the substantial reduction of industrial functions and a

TABLE 1.2 METROPOLITAN AND NON-METROPOLITAN CHANGES IN WESTERN EUROPE: 1950-75

| | Population data for 1950, 1960, 1970 and 1975 | | | | | | | |
|---|---|---|---|---|---|---|---|---|
| | 1950 | | 1960 | | 1970 | | 1975 | |
| Areal unit | Total (thousands) | % of total | Total (thousands) | % of total | Total (thousands) | % of total | Total (thousands) | % of total |
| Core | 75 314 | 38.20 | 85 261 | 40.17 | 94 479 | 40.86 | 95 190 | 40.18 |
| Ring | 88 441 | 44.85 | 93 351 | 43.98 | 103 368 | 44.71 | 109 093 | 46.06 |
| Non-Metropolitan | 33 426 | 16.95 | 33 635 | 15.85 | 33 345 | 14.42 | 32 582 | 13.76 |
| Total | 197 181 | 100.00 | 212 248 | 100.00 | 231 192 | 100.00 | 236 865 | 100.00 |

| | Population change 1950-60, 1960-70 and 1970-75 a. | | | | | | | | |
|---|---|---|---|---|---|---|---|---|---|
| | 1950-60 | | | 1960-70 | | | 1970-75 | | |
| Areal unit | Absolute change (thousands) | % change | % of total | Absolute change (thousands) | % change | % of total | Absolute change (thousands) | % change | % of total |
| Core | 9 947 | 13.21 | 66.02 | 9 218 | 10.81 | 48.66 | 711 | 0.75 | 12.53 |
| Ring | 4 910 | 5.55 | 32.59 | 10 017 | 10.73 | 52.88 | 5 725 | 5.54 | 100.92 |
| Non-Metropolitan | 209 | 0.63 | 1.39 | -290 | -0.87 | -1.53 | -763 | -2.28 | -13.44 |
| Total | 15 067 | 7.64 | 100.00 | 18 944 | 8.93 | 100.00 | 5 673 | 2.45 | 100.00 |

Note: a. Nine country region: Britain, Sweden, Norway, Denmark, France, Belgium, The Netherlands, Spain, Italy. (Separate core and ring figures not available for German Federal Republic)

Source: Peter Hall and Dennis Hay, Growth Centers in the European Urban System (University of California Press, Berkeley, California, 1980), p.86

corresponding increase in office and other service activities.[18] Consequently, a variety of suburban developments were employed to accommodate the residential and eventually industrial and commercial overspill. While the patterns varied from housing estates with single family homes outside of English cities, to large scale multi-story units surrounding cities in such countries as France, Israel, Japan, the Netherlands, USSR and Sweden, population dispersion to outlying suburban areas or new towns became a conscious public policy in most industrialized nations.[19] Together with such policies, individual preferences for outlying residential areas, and the dramatic rise in auto ownership throughout the industrialized world after 1950, metropolitan decentralization was further encouraged (see Table 1.3).[20] At the same time, central cities began to lose population. As Hall and Hay observed:

> . . . by the 1960s, a reversal had taken place: though metropolitan areas were still growing, they were decentralizing people from cores to rings. After 1970, this process accelerated, so that the cores virtually ceased to grow - with continuing losses from the non-metropolitan areas - the rings actually accounted for more than the entire net gain of the population.[21]

Table 1.2 shows that the growth of European urban cores declined from two-thirds of net population change during the 1950s, to less than one-half in the 1960s, and to slightly more than one-tenth during the 1970-75 period. Again, other industrialized nations, such as Japan, followed a similar sequence where central cities like Tokyo actually experienced net out-migration since the mid-1960s (see Table 1.4).[22] In his study of the Japanese urban system, Chauncy Harris noted:

> As the Tokyo metropolitan area expanded successive censuses showed that the zone of most rapid growth moved outward. In the period 1970-75, the most rapid growth took place in the band 31-40 kilometers from the center of Tokyo, with slower growth inward toward the center of the city and outward toward the fringes.[23]

By 1970 the industrialized world had become overwhelmingly metropolitan in character (88 percent

TABLE 1.3 LEVEL OF CAR OWNERSHIP CHANGES IN SELECTED COUNTRIES: 1955-85

| | Passenger Cars per 1000 Population | | | | | | |
|---|---|---|---|---|---|---|---|
| Country | 1955 | 1960 | 1965 | 1970 | 1975 | 1980[a] | 1985[a] |
| Austria | 10 | 55 | 107 | 162 | 227 | 315 | |
| Belgium | 56 | 82 | 135 | 213 | 255 | 303 | |
| Denmark | 57 | 106 | 181 | 260 | 313 | 377 | 439 |
| Finland | 20 | 42 | 99 | 146 | 180 | 221 | |
| France | 74 | 118 | 185 | 246 | 298 | 352 | 415 |
| Germany | 32 | 81 | 157 | 230 | 276 | 297 | |
| Greece | 2 | 6 | 12 | 26 | 46 | 101 | |
| Iceland | 63 | 75 | 140 | 185 | 280 | 344 | |
| Ireland | 46 | 61 | 97 | 130 | 182 | 227 | |
| Italy | 18 | 40 | 103 | 188 | 243 | 297 | 348 |
| Luxembourg | 71 | 118 | 185 | 268 | | | |
| Netherlands | 25 | 45 | 103 | 192 | 269 | 333 | 364 |
| Norway | 36 | 62 | 126 | 191 | 283 | 346 | 388 |
| Portugal | 12 | 18 | 26 | 47 | 75 | 110 | |
| Spain | 5 | 10 | 25 | 71 | 163 | 258 | |
| Sweden | 87 | 159 | 233 | 286 | 380 | 450 | 500 |
| Switzerland | 56 | 94 | 155 | 221 | 267 | 310 | |
| UK | 71 | 108 | 168 | 222 | 290 | 352 | 387 |
| USA | 315 | 341 | 387 | 433 | 460 | 478 | 490 |

Note: a. estimated

Source: Ralph Gakenheimer (ed.), The Automobile and the Environment: An International Perspective (MIT Press, Cambridge, Mass., 1978), p. 486

TABLE 1.4 CHANGES IN NET MIGRATION IN CENTRAL AND SUBURBAN TOKYO: 1959-75

| Year | To Suburban Tokyo (thousands) | | | To Central Tokyo (thousands) | | | Total in-migration from Non-Metro (thousands) |
|---|---|---|---|---|---|---|---|
| | From Center | From Non-Metro | Total | From Suburbs | From Non-Metro | Total | |
| 1959 | 33 | 63 | 96 | -33 | 259 | 226 | 322 |
| 1960 | 50 | 88 | 138 | -50 | 268 | 218 | 356 |
| 1961 | 72 | 112 | 184 | -72 | 265 | 193 | 377 |
| 1962 | 124 | 129 | 253 | -124 | 259 | 135 | 388 |
| 1963 | 160 | 129 | 289 | -160 | 249 | 89 | 378 |
| 1964 | 180 | 138 | 318 | -180 | 218 | 38 | 356 |
| 1965 | 176 | 118 | 294 | -176 | 206 | 30 | 324 |
| 1966 | 191 | 97 | 288 | -191 | 197 | 6 | 294 |
| 1967 | 182 | 110 | 292 | -182 | 170 | -12 | 280 |
| 1968 | 187 | 119 | 306 | -187 | 165 | -22 | 284 |
| 1969 | 197 | 114 | 311 | -197 | 155 | -42 | 269 |
| 1970 | 217 | 130 | 347 | -217 | 141 | -76 | 271 |
| 1971 | 196 | 106 | 302 | -196 | 126 | -70 | 232 |
| 1972 | 206 | 74 | 280 | -206 | 106 | -100 | 180 |
| 1973 | 229 | 44 | 273 | -229 | 71 | -158 | 115 |
| 1974 | 192 | 26 | 218 | -192 | 47 | -145 | 73 |
| 1975 | 165 | 20 | 185 | -165 | 45 | -120 | 65 |

Note: Central Tokyo includes the 23 wards of Tokyo prefecture and the city of Yokahama; Suburban Tokyo includes the rest of Tokyo prefecture, Kanagawa prefecture (excluding Yokahama city), and all of Saitama and Chiba prefectures. Non-metropolitan Regions include all of Japan excluding metropolitan Tokyo.

Source: Toshio Kuroda, 'The Impact of Internal Migration on the Tokyo Metropolitan Region', in J. W. White (ed.), The Urban Impact of Internal Migration (Institute for Research in Social Science, University of North Carolina, Chapel Hill, North Carolina, 1979), p.44

of the total population in Europe, 72 percent in Japan, and 69 percent in the United States) and massive suburbanization represented the lion's share of national population growth.[24] In Europe, for example, a vast network of metropolitan regions emerged by 1970 which captured the majority of 1950-70 total population growth, primarily as suburban development (see Figure 1.1).[25] Despite recent pressures to intensify centralization in metropolitan areas, such as the increase in the cost of energy and the growth of participation of women in the urban labor force, suburbia in the industrial world continued to expand vigorously during the 1970s.[26] By 1975 Western European suburban population accounted for nearly 60 percent of the metropolitan residents and nearly half of the continent's total population (see Table 1.2); and by 1980, the suburban residents in the United States comprised about 60 percent of the metropolitan inhabitants and well over 40 percent of the national population (see Table 1.5). Thus, many metropolitan centers became surrounded with a growing constellation of what Robert C. Wood called 'Miniature Republics', separate communities which tried to make themselves independent and safe from other communities, especially the central city.[27] As such, many suburbs differ from one another with respect to socio-economic characteristics of the population, housing stock, public services, and non-residential activity. With the decentralization of some central city industrial and commercial functions, to varying degrees the suburbs have become urbanized, somewhat like the central city itself.[28] As the decentralization of population and economic functions has increased within metropolitan areas, suburbs have come under increasing pressure to serve a wider range of social and economic activities than the primarily residential function originally envisioned.[29]

However, during the 1970s, some of the most industrialized and urbanized nations, such as Australia, Canada, and the United States, began to experience population and economic dispersion that spilled out well beyond suburbia in the largest metropolitan areas down the urban hierarchy to smaller urban regions, and out to less-developed rural areas.[30] As Tables 1.6 and 1.7 reveal, in the United States and Britain during the 1970s, small to moderate-sized metropolitan areas expanded more rapidly than the largest ones; and in the United States non-metropolitan regions grew substantially

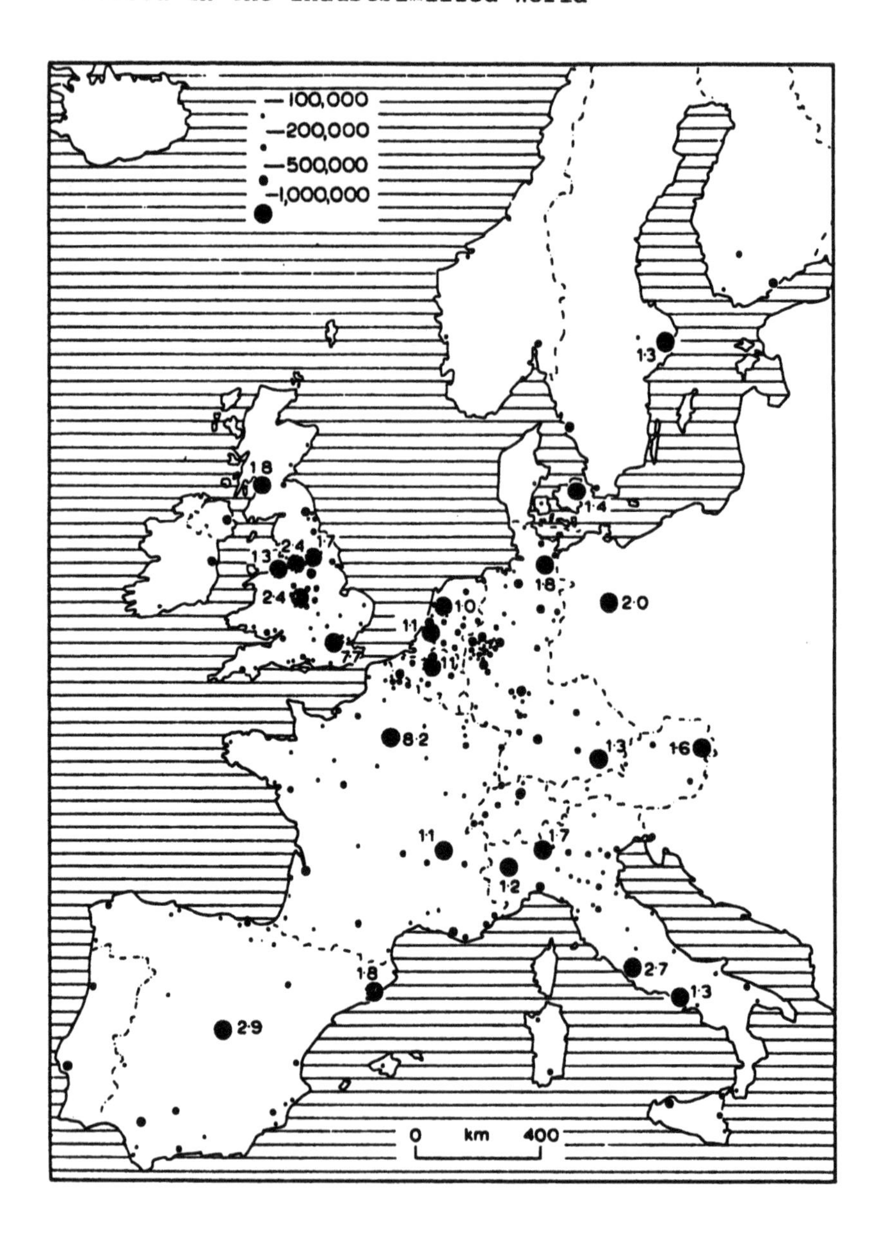


FIGURE 1.1 METROPOLITAN AREAS OF MORE THAN 100,000 INHABITANTS IN WESTERN EUROPE: 1970

Source: Hugh D. Clout, 'Population and Urban Growth', in Hugh D. Clout (ed.), Regional Development in Western Europe (John Wiley, New York, 1981), p.46

TABLE 1.5 PERCENTAGE DISTRIBUTION BY METROPOLITAN, NON-METROPOLITAN AND CENTRAL CITY STATUS IN THE UNITED STATES: 1910-80

| Population/Percent | 1910 | 1920 | 1930 | 1940 | 1950 | 1960 | 1970 | 1980 |
|---|---|---|---|---|---|---|---|---|
| Metropolitan population[a]. | 28.3 | 34.0 | 44.6 | 47.8 | 56.1 | 62.9 | 68.6 | 72.8 |
| Inside central cities | 21.7 | 25.3 | 30.8 | 32.5 | 32.8 | 32.3 | 31.4 | 29.4 |
| Outside central cities | 6.6 | 8.7 | 13.8 | 15.3 | 23.3 | 30.6 | 37.2 | 43.4 |
| Non-Metropolitan population | 71.7 | 66.0 | 55.4 | 52.2 | 43.9 | 37.1 | 21.4 | 27.2 |
| Percent of metropolitan population- | | | | | | | | |
| Inside central cities | 76.7 | 75.4 | 70.1 | 68.0 | 58.5 | 51.4 | 45.8 | 40.4 |
| Outside central cities | 23.3 | 35.6 | 30.9 | 32.0 | 41.5 | 48.6 | 54.2 | 59.6 |

a. Data for 1910 through 1940 are for metropolitan districts as officially defined at each date. Data for 1950 through 1980 are for standard metropolitan statistical areas as defined at each date.

Source: John F. Long _Population Decentralization in the United States_ (U.S. Bureau of the Census, Washington, D.C., 1981), p.65

faster than metropolitan areas as a whole. Indeed, in the United States and Canada numerous outlying 'urbanized counties' were identified which have no central city but contain all the functions and labor force of a metropolitan area in an internally dispersed pattern.[31] In fact, Table 1.6 shows that it was the smallest and most peripheral settlements in both metropolitan and non-metropolitan areas which grew fastest in the United States during the 1970s. In addition, studies of urban changes in several other industrialized nations, such as the work of Vining and Kontuly, indicate that migration to the largest metropolitan areas has slowed or reversed dramatically during the 1970s, suggesting a dispersion of population to peripheral areas in a number of industrialized countries (Japan, Sweden, Italy, Norway, Denmark, New Zealand, Belgium, France, West Germany, East Germany, and the Netherlands).[32]

While some observers claim that decentralization of population and economic activity could be the result of public policies undertaken by industrialized countries since World War II in order to divert growth away from large metropolitan areas to less developed regions,[33] others argue that such decentralization is the result of market forces inherent in advanced economies.[34] Still others, while accepting the broad scale decentralization phenomenon in the United States, are not convinced that the rest of the industrialized world has replicated the American pattern beyond suburbanization; indeed, several nations have experienced the reverse - continued dominance of growing urban cores.[35]

Yet, most observers of urban systems in the industrialized world agree that some form of evolutionary pattern may be at work.[36] These concepts, originally established by such economic theorists as Albert Hirschman and Gunnar Myrdal in the late 1950s, suggest that the initial stages of development occur around one or a few favored 'growth poles'.[37] These growing areas, provided with incentives, attract great concentrations of investment, population, and resources in order to create economies of agglomeration necessary for improved efficiency in the production, distribution, and consumption of desired goods and services.

As shown on Table 1.8, a model of urban development emerges with the first stage representing 'concentration', which involves the 'polarization effects' - the growth of large urban

TABLE 1.6 POPULATION CHANGE FOR INCORPORATED PLACES BY INITIAL SIZE AND METROPOLITAN STATUS IN THE UNITED STATES: 1960-80

| Metropolitan Status and Place Size[a] | Number of Places[b] | 1980 Population (thousands)[b] | Average Annual Rate of Population Change Per 1000 Population | |
|---|---|---|---|---|
| | | | 1960-70[c] | 1970-80[d] |
| Metropolitan | 6,670 | 165,993 | 15.8 | 9.5 |
| Outside Places | - | 53,579 | 22.7 | 19.8 |
| Places Under 2,500 | 3,317 | 3,094 | 33.5 | 27.4 |
| Places Over 2,500 | 3,353 | 109,321 | 12.0 | 4.1 |
| 2,500 to 10,000 | 1,768 | 9,181 | 30.4 | 20.1 |
| 10,000 to 25,000 | 776 | 12,359 | 25.0 | 14.6 |
| 25,000 to 50,000 | 394 | 13,918 | 20.9 | 7.5 |
| 50,000 and over | 415 | 73,863 | 6.0 | -0.3 |
| Non-Metropolitan | 12,411 | 60,511 | 3.8 | 14.3 |
| Outside Places | - | 32,767 | 1.7 | 19.2 |
| Places Under 2,500 | 9,980 | 6,951 | 6.1 | 12.1 |
| Places Over 2,500 | 2,431 | 20,794 | 6.2 | 7.8 |
| 2,500 to 10,000 | 1,817 | 8,850 | 7.4 | 9.2 |
| 10,000 to 25,000 | 477 | 7,292 | 6.8 | 6.9 |
| 25,000 and over | 137 | 4,652 | 1.6 | 5.0 |

a. Metropolitan Status as of January 1, 1980
b. Place Size as of April 1, 1980
c. Place Size as of April 1, 1960
d. Place Size as of April 1, 1970

Source: John F. Long, _Population Decentralization in the United States_ (U.S. Bureau of the Census, Washington, D.C., 1981), p.72

TABLE 1.7 POPULATION CHANGE BY FUNCTIONAL REGION IN BRITAIN: 1971-81

| Functional Region | Population (thousands) | % change |
|---|---|---|
| London Metropolitan Region | | |
| Dominant region | -740 | -8.6 |
| Sub-dominant region | 303 | 7.2 |
| Provincial Conurbations | | |
| Dominant region | -517 | -8.3 |
| Metropolitan region | 116 | 2.1 |
| Other Metropolitan Regions | | |
| Dominant region | -58 | -0.8 |
| Sub-dominant regions | 379 | 5.2 |
| Freestanding regions | 1,118 | 5.8 |

Source: Gordon E. Cherry, 'Britain and the Metropolis: Urban Change and Planning in Perspective', Town Planning Review, vol. 2, no. 55 (1984), p.29

centers which dominate and drain the hinterlands of people, resources, and capital. Gradually, 'trickling down effects' result from diseconomies of scale in large urban areas (e.g., traffic congestion, overcrowding, high land costs, pollution) and new investment opportunities in other regions, and government policies to re-direct economic growth away from heavily-developed areas overtake the 'polarization effects' and a process of decentralization sets in. At first, decentralization will manifest itself with the growth and subsequent dominance of suburban rings, then with the decline of the central cities, and finally with population and economic dispersal away from the older or larger metropolitan areas to new growth poles in smaller urban regions and to outlying, less-developed areas. Despite possible transportation costs associated with a system of widespread decentralization,[38] John Long argues in his study of American settlement patterns that such decentralization provides improved functional efficiency and is the interactive result of a highly developed nation:

TABLE 1.8 A MODEL OF URBAN DEVELOPMENT STAGES

| Type | Stage | Population Change Core | Ring | Characteristics Metropolitan Area |
|---|---|---|---|---|
| 1 | Centralization | + | - | + |
| 2 | Absolute Centralization | ++ | + | ++ |
| 3 | Relative Centralization | + | ++ | + |
| 4 | Relative Decentralization | - | + | + |
| 5 | Absolute Decentralization | - | + | - |
| 6 | Decentralization | -- | - | - |

Sources: Peter Hall and Dennis Hay, Growth Centres in the European Urban System (University of California Press, Berkeley, California, 1980), pp.229-31; and Norbert Vanhove and Leo H. Klaassen, Regional Policy: A European Approach (Allanheld, Osmun, Montclair, New Jersey, 1980), pp.180-90

> First, increased societal development requires an increased intensity of social interaction to sustain high levels of production. Second, the high levels of social interaction of all kinds that occur when a population is confined within a small area leads to congestion that can be counter-productive to social interaction. Third, with advanced development a society develops more efficient interpersonal linkages through improved communication, transportation, and other adaptations of formal organizations. These improved linkages permit the channeling of social interaction so that productive activities can be performed without interference from other forms of social interaction. The society can then achieve higher levels of functional integration of all parts of society - even those at great distances. Consequently, increased societal development brings increasing pressures for deconcentration and at the same time reduces the effect of distance so that deconcentration can occur.[39]

In addition to allowing decentralization by reducing the influence of physical proximity, there is evidence that economically advanced societies demand decentralization in order to meet the increasing requirements for more space by modern industrial production processes, service, and retail activities; increasingly affluent households wanting lower density housing; public desires for a more efficient use of private motor vehicle transportation; and a growing concern about quality of life factors such as accessibility to recreational facilities and the natural environment.[40] In his international study of the metropolis, Gordon Cherry finds that in industrialized societies:

> Concentration has been replaced by dispersal, aided and abetted basically by technological developments in transport (particularly private) and new factors in employment locations, but also underpinned by public sector policies favouring reduction of densities and pursuing programmes of regional economic planning. . .[41]

What may be unfolding in advanced economies is

a process of metropolitanization of the countryside, such that social and economic differences between urban and rural regions will continue to narrow, as has occurred in the case of central cities and suburbs. Eventually, such a process (due to either market or planned activity) is likely to transform entire nations into culturally and economically integrated metropolitan societies.[42]

This model, then, suggests that nations still in the earlier phases of industrialization, such as those in southern and eastern Europe, will have most of their population growth concentrated in urban cores, while other countries in a later phase of industrial development, such as the United States, the Netherlands and Sweden, will be involved in widespread decentralization. As Hall and Hay have concluded about European urbanization trends:

> The most likely hypothesis is that the main differences are related to stage of industrial-urban evolution; and that as this proceeds, Europe will more and more follow the American-British path.[43]

Recognizing that there are and will continue to be variations in national urban patterns because of important factors as history, culture, governmental policies, and changing housing and lifestyle preferences, it seems likely that a substantial part of the urban system of the industrial world will continue to decentralize and that suburbia will continue to play a major role in this global process. As Cherry aptly put it, 'with collapsing cores, the focus of growth now resides in suburban nodes and freestanding satellites beyond'.[44]

NOTES

[1] It has been suggested that, after a lengthy era of periodic use of traditional gathering places, permanent settlements in the form of small agriculturally based hamlets emerged in the neolithic culture about ten to twelve thousand years ago. See Lewis Mumford, The City in History (Harcourt, Brace & World, Inc., New York, 1961), pp. 5-15.

[2] Ibid., p.483.

[3] A discussion of outlying settlements near ancient cities is presented in Paul Lample, Cities and Planning in the Ancient Near East (George Braziller, New York, 1968), pp.13-22; and Gideon

Sjoberg, The Preindustrial City: Past and Present (Free Press, New York, 1960), pp.25-51.

[4] Mumford, The City in History, pp.482-525.

[5] See David Ward, Cities and Immigrants: A Geography of Change in Nineteenth Century America (Oxford University Press, New York, 1971), Chapter 4; Joel A. Tarr, 'From City to Suburb: The "Moral" Influence of Transportation Technology' in Alexander B. Callow (ed.), American Urban History (Oxford University Press, New York, 1973), p.202; and Robert C. Wood, Suburbia: Its People and Their Politics (Houghton-Mifflin, Boston, 1958).

[6] For a presentation of the Anglo-American view of the garden city concept, see Walter L. Creese, The Search For Environment: The Garden City: Before and After (Yale University Press, New Haven, 1966); and Clarence Stein, Toward New Towns for America (Reinhold Publishing Corp., New York, 1956). For a discussion of the garden city influence in other industrialized countries see Jeremy D. Alden, 'Metropolitan Planning in Japan', Town Planning Review, vol. 55, no. 1 (1984), pp.55-74; Creese, Ibid., Chapter 13; and Arthur B. Gallion and Simon Eisner, The Urban Pattern (D. Van Nostrand Company, New York, 1975), pp.98-125.

[7] John C. Bollens and Henry J. Schmandt, The Metropolis: Its People, Politics, and Economic Life (Harper and Row, New York, 1975), Chapter 1; and Donald J. Bogue, Population Growth in Standard Metropolitan Areas 1900-1950 (Housing and Home Finance Agency, Washington, D.C., 1953).

[8] Peter Hall, et al., The Containment of Urban England, Vol. 1 (George Allen and Unwin, London, 1973), p.64.

[9] Mumford, The City in History, p.529.

[10] See John R. Meyer, John F. Kain, and Martin Wohl, The Urban Transportation Problem (Harvard University Press, Cambridge, Mass., 1965), Chapter 2; David C. Thorns, Suburbia (MacGibbon and Kee, London, 1972), Chapter 4.

[11] These are large scale residential developments built by Levitt and Sons near New York City and Philadelphia during the late 1940s and early 1950s. See David Poponoe, The Suburban Environment: Sweden and the United States (The University of Chicago Press, Chicago, 1977), Chapter 5.

[12] Hugh D. Clout, 'Population and Urban Growth', in Hugh D. Clout, (ed.), Regional Development in Western Europe (John Wiley, New York, 1981), pp.35-59.

[13] Peter Self, Planning the Urban Region: A Comparative Study of Policies and Organization (The University of Alabama Press, Alabama, 1982), p.38.

[14] This process continued in Western Europe so that during the 1970s the migrant workers and dependants alone totaled about 12 million people. See Clout, 'Population and Urban Growth', pp.38-51.

[15] Peter Hall and Dennis Hay, Growth Centres in the European Urban System (University of California Press, Berkeley, California, 1980), p.87.

[16] See Chauncy D. Harris, 'The Urban and Industrial Transformation of Japan', The Geographic Review, vol. 72, no. 1 (1982), pp.50-89; and Toshio Kuroda, 'The Impact of Internal Migration on the Tokyo Metropolitan Population' in James W. White (ed.), Urban Impact of Internal Migration (Institute for Research in Social Science, University of North Carolina at Chapel Hill, Chapel Hill, 1979), pp.33-52.

[17] Self, Planning the Urban Region, p.39.

[18] Clout, 'Population and Urban Growth', pp.48-50.

[19] See Ibid., pp.48-53; Ervin Y. Gallanty, New Towns: Antiquity to the Present (George Braziller, New York, 1975), Chapter 5; and Thorn, Suburbia, Chapter 4.

[20] For example, Western Europe after 1950 experienced a substantial increase in car ownership so that by 1976 about 85 million private cars were on the expanding continental road network. See Clout, Ibid., p.53; and James H. Johnson, 'Geographical Processes at the Edge of the City' in James H. Johnson (ed.), Suburban Growth, (John Wiley and Sons, New York, 1974), pp.1-16.

[21] Hall and Hay, Growth Centres, p.87.

[22] Kuroda, 'The Impact of Internal Migration', pp.43-52.

[23] Harris, 'The Urban and Industrial Transformation Of Japan', p.85.

[24] See Hall and Hay, Growth Centres, pp.85-88; Harris, Ibid., p.52; Donald N. Rothblatt, Planning the Metropolis: The Multiple Advocacy Approach (Praeger, New York, 1982), pp.2-6; and United Nations, Patterns of Urban and Rural Population Growth (United Nations, New York, 1980), Chapter 2.

[25] Hall and Hay, Growth Centres, pp.154-8.

[26] As women increased their participation in the labor force, there appeared to be a growing demand for households to be located close to the central city center in order to minimize working women's journey to work. See Donald N. Rothblatt,

Daniel J. Garr and Jo Sprague, The Suburban Environment and Women (Praeger, New York, 1979), Chapters 4 and 5.

[27] Wood, Suburbia.

[28] Louis H. Masotti, 'Suburbia Reconsidered: Myth and Counter-Myth' in Louis H. Masotti and Jeffrey K. Hadden (eds.), The Urbanization of the Suburbs (Sage, Beverly Hills, California, 1973), pp.15-22.

[29] Thorns, Suburbia, Chapter 5.

[30] See Brian J. L. Berry, 'The Counterurban Process: Urban America Since 1970' in Brian J. L. Berry (ed.), Urbanization and Counterurbanization (Sage, Beverly Hills, California, 1976), pp.17-30; Larry S. Bourne and M. I. Logan, 'Changing Urbanization Patterns at the Margin: The Examples of Australia and Canada', in Ibid., pp.111-43; John F. Long, Population Deconcentration in the United States (U.S. Bureau of the Census, Washington, D.C., 1981), Chapters 2 and 3; and Maurice Yeates, North American Urban Patterns (V. H. Winston and Sons, New York, 1980), Chapter 6.

[31] In one study, as many as 49 such urbanized counties were identified in the United States. See Mark Gottdiener, 'The New Form of Settlement Space: Conceptualizing Decentralization', Paper delivered at the Association of Collegiate Schools of Planning Conference, San Francisco, October 1983. Similar county developments in Canada were reported in Yeates, North American Urban Patterns, pp.142-4.

[32] See Daniel R. Vining, Jr. and Thomas Kontuly, 'Population Dispersal for Major Metropolitan Regions: An International Comparison', International Regional Science Review vol.3, no. 1 (1978), pp.49-73; and other individual case studies such as J. G. Borchert, 'The Dutch Settlement System' in H. Van der Haegen (ed.), Western European Settlement Systems (Instituut voor Sociale en Economische Geografie Katholike Universiteit te Lueven, Lueven, 1982), pp.207-48; and Harris, 'The Urban and Industrial Transformation of Japan', pp.88-9.

[33] James L. Sundquist Dispersing Population: What America Can Learn From Europe (The Brookings Institution, Washington, D.C., 1974); B. S. Khorev and V. M. Moiseenko, 'Urbanization and Redistribution of the Population of the U.S.S.R. in Sidney Goldstein and David S. Sly (eds.), Patterns of Urbanization: Comparative Country Studies, Vol. 2 (International Union for the Scientific Study of

Population, Dolhain, Belgium, 1977), pp.634-720.

[34] See Gordon E. Cherry, 'Britain and the Metropolis: Urban Change and Planning in Perspective', Town Planning Review, vol. 55, no. 1 (1984), pp.5-53; and Long, Population Decentralization in the United States, Chapters 1 and 5.

[35] Hall and Hay, Growth Centres, Chapter 7.

[36] For example, see Cherry, 'Britain and the Metropolis', pp.31-2; Hall and Hay, Ibid.; Harris, 'Urban and Industrial Transformation of Japan', pp.88-9; Long, Population Decentralization in the United States, Chapter 5; Norbert Vanhove and Leo H. Klaassen, Regional Policy: A European Approach (Allenheld, Osmun & Co., Monclair, N.J., 1980), Chapter 5; and Vining and Kontuly, 'Population Dispersal for Major Metropolitan Regions', pp.68-9.

[37] Albert O. Hirschman, The Strategy of Economic Development (Yale University Press, New Haven, 1958), Chapter 10; and Gunnar Myrdal, Economic Theory and Underdeveloped Regions (G. Duckworth & Co. Ltd., London, 1957), Chapters 3-5.

[38] This refers to the well known trade-offs between transport and land consumption costs with distance from major market centers. See William Alonso, 'Location Theory' in John Friedmann and William Alonso (eds.), Regional Policy: Readings in Theory and Applications (MIT Press, Cambridge, Mass., 1975), pp.35-163; and Steven E. Plant, 'The Economics of Population Dispersal', Urban Studies, vol. 20, no. 3 (August 1983), pp.353-7.

[39] Long, Population Decentralization in the United States, pp.86-7.

[40] Berry, 'The Counterurbanization Process', pp.28-30; and Long, Ibid., Chapter 5.

[41] Cherry, 'Britain and the Metropolis', p.31.

[42] See Khorev and Moiseenko, 'Urbanization and Redistribution of the Population of the U.S.S.R.', pp.643-720; Charles L. Leven, 'Regional Variations in Metropolitan Growth and Development' in Victor L. Arnold (ed.), Alternatives to Confrontation: A National Policy Toward Regional Change (Lexington Books, Lexington, Mass., 1980), pp.329-43; and John Oosterbaan, Population Dispersal: A National Imperative (Lexington Books, Lexington, Mass., 1980), Chapter 12.

[43] Hall and Hay, Growth Centres, p.231.

[44] Cherry, 'Britain and the Metropolis', p.31.

# CHAPTER 2

# SUBURBAN TRENDS IN THREE COUNTRIES

This chapter will discuss the particular circumstances which created the pressures for suburban growth in the United States, Israel and the Netherlands. While each society experienced diffferent circumstances resulting from the tumultuous era of the Great Depression and the Second World War, in each case the adjustment to post-1945 conditions required greater levels of public sector involvement than ever before.

As pluralistic nations with traditions of participatory democracy, the countries selected for study provide a useful basis for this comparative cross-national study. In addition, as advanced capitalist economies in transition from industrial to post-industrial service societies, trends in their respective urban development patterns are similar, a feature that further broadens the base of this study for students of post-1945 metropolitan development.

## A. THE UNITED STATES: THE INSTITUTIONAL BASIS OF AMERICAN SUBURBANIZATION AFTER 1945

By the last quarter of the nineteenth century, suburbs in the United States had completed nearly a generation of rapid growth on the periphery of that nation's largest cities. American metropolitan decentralization was initially facilitated by major transportation innovations, by the informal cooperation of myriad small builders, subdividers, municipal officials and financial middle-men, by the demographic pressures of central city population growth, and by the values of a rapidly emerging middle class anchored to the arcadian hearth as a safe harbor amidst the uncertainties of competitive capitalism.[1] This synchronism of technology, social stratification, and informal urban development processes contributed decisively to the United States becoming a metropolitan nation by the year 1910. It was at the end of the first decade of the twentieth century that the United States Bureau of the Census introduced the concept of 'metropolitan district', a category of population enumeration necessitated by the growth of large cities and their more rapidly expanding environs.[2]

Until the advent of mass transit at a price affordable by the middle class, American cities were the domain of increasingly congested pedestrian traffic and primitive short-haul transportation, the efficiency of which was inversely proportional to its olfactory amenity. Few may realize that the practical extent of city size approximated that of European urbanizations during medieval times, a constraint imposed by the proximity necessary for productive interactions. Choking densities were an inevitable by-product of urban life in the late nineteenth century and approached 10,000 persons per square mile in typical American cities, but the concentrations were most severe in the largest centers.[3] For example, the five most congested wards on Manhattan's Lower East Side were literally bursting with over 200,000 inhabitants per square mile by the 1890s.[4]

However, by the onset of the First World War, American urban centers had broken the grip of the 'walking city' and population redistribution outside central cities proceeded as rapidly as transit facilities permitted. By 1930, travel by auto, truck, and bus eliminated the confines imposed by fixed-rail transit, a development which opened up a commuter-shed of as much as 2,000 square miles around urban centers. This expansion was made possible by large public investments in streets and highways.[5]

In 1945, following almost two decades of wartime sacrifice and the deprivations imposed by the Great Depression, American consumer demand exploded. Fearful of a re-emergence of the economic woes of the 1930s, the American public sector took steps to insure that the millions of returning veterans would find employment. This was partially accomplished by accelerating the transition from wartime to peacetime manufacturing, especially for automobiles, and by an unprecedented highway building program which brought new suburban areas within reach of the growth potential opened by earlier short-distance transportation innovations.[6]

Without mass access to automobile ownership, suburbanization could never have occurred in the United States. Dependence on fixed or light rail service required an economic return that would be impossible to achieve under conditions defined by low residential densities. The fixed costs imposed by commuter and/or freight rights-of-way required heavy utilization and service was inherently limited to only those points along a right-of-way. Only in areas with high population concentrations could rail lines produce a viable economic return. And because of the traffic volume required, there was a limit on the number of rights-of-way that could be profitably operated. As a result, a vast amount of territory remained inaccessible to rail transit.[7]

Between 1945 and 1970 automobile ownership became a permeating factor in American society. In 1945, 26 million automobiles were registered in the United States; in other words, nearly 25 percent of those of driving age (16 and over) could be matched with a car. By 1970, the number of autos more than trebled to 89 million, so that the ratio of population 16 and over to registered vehicles dropped to one auto for every 1.4 individuals. In addition, purchasers of vehicles were not impeded by burdensome regulations and high taxes at the federal level. Further, state and local sales levies were modest (if any existed) and impediments to obtaining a driver's license were limited to a vision test, a simple written exam, and a nominal road examination. While these variables may differ elsewhere in the world, in no case would they approach the unfettered accessibility enjoyed in the United States. Even into the 1980s, some states, including California, did not require operators of motor vehicles to carry insurance! Lastly, in the United States the price of gasoline has remained a fraction of the cost encountered in other industrialized nations. Yet, despite the escalating energy costs associated with the mid-1970s, suburbanization has continued unfettered in the United States. Indeed, during 1975-78 the population growth of counties at the metropolitan fringe increased at a rate one-third higher than that for 1970-75.[8]

The transition to vehicular transportation proved to be a prime cause of the deconcentration of jobs and industry, a process that began before World War II. For example, between 1939 and 1947 the number of workers employed in manufacturing increased in suburban areas at a rate more than twice that in the central city, and by 1958, manufacturing employment showed an even more dramatic differential.[9] While the core areas witnessed an overall loss of 6.8 percent, the metropolitan peripheries registered a 40 percent gain.[10]

There are two major reasons for the dispersion of industry. First, a twofold change occurred in vehicular transport. In addition to widespread auto ownership that enabled workers to leave transit-oriented labor markets behind, the development of trucking freed manufacturing from proximity to traditional transport entrepôts: ports and railheads. Secondly, in response to assembly line technology and changes in the handling of materials, factories expanded in a horizontal direction, an imperative that could only be accommodated by larger and less-expensive suburban parcels of land. In addition, urban congestion precluded lateral extensions of manufacturing even if the remote possibility of reasonable land costs prevailed.[11] And once industrial decentralization began to occur, economies of scale, the creation of satellite and complementary employment, and the inevitable attraction of increased population served to catapult suburbs into self-reinforcing patterns of economic and demographic expansion.[12]

These same factors exerted an even more dramatic impact on retail sales. Between 1945 and 1960 more than 2,500 suburban shopping centers were constructed, a trend which has been devastating for central city retail trade (see Table 2.1).[13] Between 1954 and 1967 the number of retail establishments declined precipitously in the central city as did the volume of sales. The distinct downtown orientation of retail activity gave way to increasingly large regional centers by the early 1960s, a trend that was accompanied by the emergence of specialized developers capable of creating a standardized shopping center format that would appeal to what has been termed the 'vast monolithic middle-class market'.[14] By the 1980s, the shopping center industry has diversified so that a range of configurations, marketing approaches, and locational elements are now offered.[15]

Of course, demographic changes provided the magnet that pulled other activities into their centrifugal flight. As Chapter 1 has demonstrated, the decentralization of population away from the urban core has defined the focus of both market processes and public policy. The situation in the

TABLE 2.1 PERCENT CHANGE IN ESTABLISHMENTS AND SALES IN SELECTED PORTIONS OF METROPOLITAN AREAS IN THE UNITED STATES: 1954-67

| | CBD | | City | | Outside City | | SMSA | |
|---|---|---|---|---|---|---|---|---|
| Size of Area | Est. | Sales | Est. | Sales | Est. | Sales | Est. | Sales |
| Total Sales | | | | | | | | |
| Total | -32.4 | 2.6 | 18.7 | 44.0 | 37.7 | 167.1 | 3.3 | 89.7 |
| 3,000,000 + | -26.0 | 12.1 | -26.3 | 34.3 | 29.9 | 132.2 | -7.4 | 70.4 |
| 1,000,000 to 3,000,000 | -26.9 | 8.3 | -23.7 | 26.8 | 30.3 | 175.0 | 3.6 | 97.6 |
| 500,000 to 1,000,000 | -38.2 | -3.7 | -8.4 | 58.4 | 51.3 | 209.0 | 14.1 | 104.0 |
| 250,000 to 500,000 | -37.6 | -6.7 | -8.6 | 61.4 | 48.0 | 193.1 | 13.4 | 104.2 |
| General Merchandise | | | | | | | | |
| Total | -22.1 | 5.5 | -21.0 | 76.4 | 40.0 | 470.6 | 2.0 | 165.6 |
| 3,000,000 + | -16.7 | 12.9 | -32.5 | 68.1 | 16.2 | 395.0 | -16.9 | 158.2 |
| 1,000,000 to 3,000,000 | -11.7 | 8.2 | -25.0 | 29.1 | 36.9 | 467.0 | 9.6 | 153.5 |
| 500,000 to 1,000,000 | -28.3 | -1.5 | -8.0 | 95.3 | 47.8 | 707.8 | 17.0 | 177.2 |
| 250,000 to 500,000 | -26.1 | 2.0 | -2.8 | 118.6 | 41.4 | 486.6 | 17.1 | 181.0 |
| Apparel | | | | | | | | |
| Total | -39.8 | -7.6 | -24.5 | 18.5 | 23.1 | 133.6 | -8.5 | 49.1 |
| 3,000,000 + | -32.2 | -1.4 | -23.7 | 14.5 | 7.3 | 105.6 | -14.7 | 37.5 |
| 1,000,000 to 3,000,000 | -34.8 | 4.5 | -34.0 | 12.8 | 25.8 | 148.2 | -6.3 | 64.8 |
| 500,000 to 1,000,000 | -46.6 | -15.3 | -19.8 | 30.6 | 35.0 | 138.5 | -1.2 | 55.3 |
| 250,000 to 500,000 | -45.2 | -17.8 | -23.0 | 21.4 | 45.6 | 192.0 | -0.8 | 56.8 |

| | | | | | | | | |
|---|---|---|---|---|---|---|---|---|
| Furniture | | | | | | | | |
| Total | -39.8 | -10.6 | -18.7 | 19.3 | 38.6 | 165.9 | 3.5 | 65.2 |
| 3,000,000 + | -45.1 | -11.1 | -25.2 | -5.7 | 12.9 | 146.5 | -11.2 | 43.0 |
| 1,000,000 to 3,000,000 | -32.8 | -3.4 | -27.5 | 11.6 | 44.3 | 169.1 | 8.7 | 77.3 |
| 5000,000 to 1,000,000 | -42.0 | -8.8 | -8.3 | 48.5 | 72.3 | 183.1 | 19.9 | 84.0 |
| 250,000 to 500,000 | -36.7 | -16.6 | -9.3 | 45.9 | 51.7 | 198.8 | 12.3 | 81.4 |

Source: United States Census data compiled by Basil G. Zimmer in 'The Urban Centrifugal Drift', in Amos H. Hawley and Vincent P. Rock (eds.), Metropolitan America in Contemporary Perspective (Halstead Press/John Wiley & Sons, New York & London, 1975), p.61

TABLE 2.2 POPULATION CHANGES FOR SELECTED URBANIZED AREAS IN THE UNITED STATES 1950-80

| | Total Population (thousands) | | | | Decennial Growth Rates (%) | | |
|---|---|---|---|---|---|---|---|
| Urbanized Areas | 1950 | 1960 | 1970 | 1980 | 1950-60 | 1960-70 | 1970-80 |
| Los Angeles - Long Beach, CA | | | | | | | |
| Cities | 2,221 | 2,823 | 3,620 | 3,328 | +27 | +12 | -8 |
| Ring | 2,147 | 3,666 | 4,730 | 6,151 | +70 | +29 | +30 |
| Total | 4,368 | 6,489 | 8,350 | 9,479 | +48 | +28 | +14 |
| New York - Northeast New Jersey | | | | | | | |
| Cities | 8,891 | 8,743 | 8,820 | 7,624 | -2 | +1 | -14 |
| Ring | 3,405 | 5,372 | 7,387 | 7,966 | +58 | +38 | +8 |
| Total | 12,296 | 14,115 | 16,207 | 15,580 | +15 | +15 | -4 |
| San Francisco - Oakland, CA | | | | | | | |
| Cities | 1,160 | 1,158 | 1,144 | 1,108 | 0 | -1 | -11 |
| Ring | 862 | 1,273 | 1,844 | 2,172 | +48 | +45 | +18 |
| Total | 2,022 | 2,431 | 2,988 | 3,190 | +20 | +23 | +1 |
| San Jose, CA | | | | | | | |
| City | 96 | 205 | 460 | 628 | +114 | +124 | +34 |
| Ring | 196 | 437 | 605 | 667 | +122 | +38 | +10 |
| Total | 292 | 624 | 1,065 | 1,295 | +120 | +66 | +22 |

| | | | | | | | |
|---|---|---|---|---|---|---|---|
| Chicago, IL | | | | | | | |
| City | 3,897 | 3,898 | 3,366 | 3,005 | 0 | -14 | -11 |
| Ring | 1,024 | 2,061 | 3,017 | 3,775 | +101 | +46 | +25 |
| Total | 4,921 | 5,959 | 6,383 | 6,780 | +21 | +7 | +7 |
| Philadelphia - Southwest New Jersey | | | | | | | |
| Cities | 2,072 | 2,003 | 1,949 | 1,688 | -3 | -3 | -13 |
| Ring | 851 | 1,633 | 2,073 | 2,425 | +92 | +26 | +16 |
| Total | 2,953 | 3,636 | 4,022 | 4,113 | +23 | +11 | +2 |
| Houston, TX | | | | | | | |
| City | 586 | 938 | 1,233 | 1,595 | +57 | +31 | +29 |
| Ring | 104 | 202 | 445 | 818 | +93 | +120 | +83 |
| Total | 690 | 1,140 | 1,678 | 2,413 | +65 | +47 | +44 |
| Detroit, MI | | | | | | | |
| City | 1,850 | 1,670 | 1,511 | 1,203 | -10 | -10 | -20 |
| Ring | 810 | 1,867 | 2,475 | 2,606 | +131 | +33 | +5 |
| Total | 2,660 | 3,537 | 3,986 | 3,809 | +33 | +13 | -4 |
| Boston, MA | | | | | | | |
| City | 801 | 697 | 641 | 563 | -13 | -8 | -12 |
| Ring | 1,433 | 1,716 | 2,015 | 2,116 | +20 | +17 | +5 |
| Total | 2,234 | 2,413 | 2,656 | 2,679 | +8 | +19 | +1 |

Source: U.S. Bureau of the Census, Census of Population: 1950, vol. 2, table 33; Census of Population: 1960, vol. 1, table 20; Census of Population: 1970, vol. 1, table 24 (Government Printing Office, Washington, D.C., various); U.S. Bureau of the Census, Population and Land Areas of Urbanized Areas for the United States and Puerto Rico, 1980 & 1970. Supplementary Report, Table 7, (Government Printing Office, Washington, D.C., 1984)

United States is illustrated by the relative rates of population growth for a variety of selected metropolitan areas (See Table 2.2). In every case but one, the suburban ring has far outstripped population growth in the central cities by spectacular margins. Even when metropolitan area population has declined, as is the case with New York and Detroit between 1970 and 1980, suburban rings continued to increase their number of residents. A significant exception to this pattern is the case of the San Jose metropolitan area which has witnessed higher rates of growth for the city than for its surrounding communities during the 1960-70 and 1970-80 decades. The explanation for this discrepancy lies in the fact that annexation has been an important factor in the growth of many American cities, primarily those in the South and West. Urban centers such as Oklahoma City, Houston, Phoenix, Dallas, San Diego, Corpus Christi and San Antonio, Texas, and San Jose, all annexed at least one hundred square miles between 1950 and 1972.[16] In particular, the actions of San Jose during those years stimulated a round of geopolitical warfare within Santa Clara County which resulted in the 'defensive formation' of new municipalities;[17] one of these, Cupertino (incorporated in 1955), will be discussed in subsequent chapters (see Table 2.4).

A major corollary of the decentralization of population is the changing geography of poverty. Within metropolitan areas, concentrations of the poor, particularly those of racial minorities, have become an essential characteristic of many American central cities. For example, between 1970 and 1979, Blacks increased their share of central city population from 16.4 percent to 23.4 percent.[18] These figures represent an inevitable result of five sets of forces: first, the mechanization of agriculture triggered a flow of population to larger urban centers; second, migration occurred from relatively poor regions, such as the South, Puerto Rico, and Appalachia, to more prosperous ones, e.g., the Pacific Coast and the Northeast and North Central regions of the United States; third, central cities possessed the largest concentrations of older and inexpensive housing units, most of which enjoyed good access to employment as well as to public transportation; fourth, racial prejudice has denied low-income minorities access to suburban housing opportunities that poor white families are far more likely to enjoy;[19] and fifth, public policy has provided inducements to the middle classes to move to the suburbs which are either beyond the reach of low-income families (favorable home financing and income tax policy and the high price of auto-based suburban transportation) or designed to discourage their penetration of the suburban ring. For example, Jackson has demonstrated that criteria established by the Federal Housing Administration (FHA) during the 1930s resulted in active discouragement to racial minority groups seeking to buy suburban housing.[20] This practice was also observed by Gans in the formative years of Levittown, Pennsylvania, a community in which he had taken up residence in order to record first-hand observations of life in a new suburban community.[21] The racial polarization which has occurred in the United States has been the source of a circular pattern of tensions that has increased the impetus of middle and upper-class whites to leave for the suburbs. The civil disturbances of the mid-1960s were a reflection of this division in American society and there is little evidence that their causes have been addressed.[22]

However, as in other nations, suburbanization in the United States was in no way the sole domain of the impersonal forces which determined the direction of a free market economy. In contrast to the *ad hoc*, diffused, and fragmented nature of suburban development in the years prior to the Second World War, its encouragement, if not management, in the years after 1945 was often just the opposite: centralized, federalized, institutionalized.[23] Despite that volume of scholarly prose concerning the balkanization of local government and the inefficiency of the construction industry, suburban development has greatly been determined by decisions rendered in Washington, D.C.[24]

Federal aid for highway construction began in 1916 and increased in virtually every succeeding year, exceeding $4 billion per annum during the 1960s. In another manner of speaking, highway building enjoyed a half-century head start on mass transit funding from the federal level, an event which did not occur until the Urban Mass Transportation Act of 1964. The creation of the Highway Trust Fund in 1956 established a method for financing a new interstate highway system out of revenues derived from excise taxes on vehicles, gasoline, and tires. With funding requiring only a 10 percent local share to match the 90 percent balance derived from federal sources, municipal, county, and state government officials were more than eager to participate in the economic and political benefits accruing from large scale highway-building projects. They opened up broad expanses of territory to urban development as circumferential freeways inexorably began to define a metropolitan realm predicated on automobile ownership.

With vast quantities of inexpensive land made acccessible by these highways, the federal government thus eliminated a major constraint on supplying housing in outlying areas. In the late 1940s, it had solved one of the major problems on the demand side of the equation by infusing the Depression-era Federal Housing Administration (FHA) mortgage insurance programs with unprecedented appropriations which, like highway funding, would continue to escalate in magnitude. Similarly, Veterans' Administration (VA) loan guarantees compounded the resources that channeled the home-building industry to suburban locations. Even before World War II, the thrust of these mortgage insurance activities was overwhelmingly suburban; however, after 1945, they escalated dramatically.[25]

The mechanics of VA/FHA mortgage insurance made it possible to own a family-sized, new suburban house for less than the cost of renting a central city apartment. This was accomplished by long-term (twenty to thirty year) loan amortization, a significant change from pre-Depression financing techniques of short-term (three to eight year), interest only, balloon payment mortgage instruments. Further, interest rates on these government-insured loans were somewhat below the market rate because of the reduced risk resulting from government backing. In addition, down payments of five percent or less were customary, making home ownership an easily attained goal for almost any middle class American family.

With highway projects making accessible land available on an unprecedented scale, and inexpensive financing insured by federal guarantees, even the houses themselves bore the stamp of the national government. Federal involvement in suburban residential growth was specifically geared to accommodate the large builder who was instrumental in the formulation of FHA construction standards, the indispensable prerequisite for mortgage insurance.[26] As a case in point, Levitt and Sons boasted the largest line of credit ever provided to a privately-owned builder, a direct result of the organization's being able to secure FHA production advances and commitments to finance four thousand homes before even a single lot was cleared. The success of the Levitt colossus was exemplified by its product, a well-designed suburban dream house for as little as $56 per month.[27]

In addition, the Levitts applied vertical organization and rationalization to homebuilding in a manner more reminiscent of the automobile assembly line than of the traditional building trades. By 1948, 35 houses per day were completed in Levittown, Long Island, a community situated on former potato fields thirty miles east of New York City that would eventually comprise over ten thousand homes. Not only were Levitt's construction and financing practices refined to the highest levels of precision, but marketing techniques made it possible to process as many as 350 buyers in one day with all legal, financial, and real estate details handled for only $10 in closing costs.[28] By 1949, large builders such as Levitt had accounted for one-quarter of all new single-family housing units and for nearly two-thirds of all such units by 1959.[29]

However, the permeating federal presence was not limited to only housing and highways. Water and sewage projects provided necessary infrastructure and services. The onset of the Cold War boosted defense spending and construction of military bases. Corporate contractors to the military prospered. These developments added still another impetus to suburbanization, particularly in the sunbelt states. Of the nearly 500,000 federal jobs created between 1960 and 1975, 27 percent were in the West, a generous allocation given that region's 15 percent share of the national population.[30]

Perhaps no region has profited more from the guns and butter federal presence than has the San Francisco Bay Area (see Figure 2.1). Established as a center of war-related manufacturing during the 1940s, its position as a munitions and research center was further enhanced during the Korean War. In addition, the presence of two major research universities (the University of California-Berkeley, and Stanford University), and the Moffett Field Naval Air Station and other major military installations, all channeled unprecedented flows of federal funds to the region. Within a decade of the cessation of hostilities in Korea, Bay Area scientist/entrepreneurs, paticularly those in Santa Clara County, had captured a major share of the

FIGURE 2.1: THE SAN FRANCISCO BAY REGION

Source: 1970 USGS - San Francisco Bay Region Map.

growth generated by the innovative military and research applications of what is now called 'high tech'. As a magnet for migrants, Santa Clara County has attracted a highly educated and professionalized work force pyramided atop a substantial pool of indispensable lower-skilled production workers.[31] In the 1950-60 decade, the County more than doubled its population, a rate of growth which exceeded that of the state of California by more than a factor of two and that of the nation as a whole by more than six.

The suburbanization and industrialization of the Santa Clara Valley since 1940 transformed one of the fifteen most productive agricultural counties in the United States into one of the most technologically-advanced in the world. Known as 'Silicon Valley' since the early 1970s, the County added more than one million residents by 1980 during four decades of explosive population growth (see Table 2.3).

In 1940 there were nine incorporated jurisdictions in the County, of which San Jose, with 68,500 residents, was the largest. It served as the county seat and the hub of agriculture, with many canneries, food machinery industries, and supportive businesses and services. The other towns, with one key exception, were the scattered service centers for 100,000 acres in orchards and 8,000 in vegetables. 'The urban half of the population', former County Planning Director Karl Belser recalled, 'was the exact counterpart of the farm community', for their relationship was deeply intertwined. The towns, Belser continued,

> provided the financial, retail, professional and personal services. They were also the market for some of the farm produce. However, the key to the economic life was the joint activity where the produce of the farm was processed and prepared for delivery to the world market. Each dollar of value produced on the farm was recycled through the economy several times.[32]

The one atypical community, Palo Alto, would be the point of genesis for the great changes that were to engulf the agrarian world that had existed in Santa Clara County until the 1940s. Palo Alto, 20 miles northwest of San Jose, grew up around Leland Stanford's great university, which had been established in 1885. In 1940 it had less than 17,000 residents, most of whom were affiliated with the

TABLE 2.3 POPULATION OF CITIES AND UNINCORPORATED AREAS IN SANTA CLARA COUNTY: 1920-80

| | 1920 | 1930 | 1940 | 1950 | 1960 | 1970 | 1980 |
|---|---|---|---|---|---|---|---|
| Alviso | 500 | 400 | 700 | 700 | 1,174 | a. | a. |
| Campbell | - | - | - | - | 11,863 | 23,797 | 27,067 |
| Cupertino | - | - | - | - | 3,664 | 17,895 | 34,420 |
| Gilroy | 2,900 | 3,500 | 3,600 | 4,900 | 7,348 | 12,684 | 21,641 |
| Los Altos | - | - | - | - | 19,696 | 25,062 | 25,769 |
| Los Altos Hills | - | - | - | - | 3,412 | 6,871 | 7,421 |
| Los Gatos | 2,300 | 3,200 | 3,600 | 4,900 | 9,036 | 22,613 | 26,593 |
| Milpitas | - | - | - | - | 6,572 | 26,561 | 37,820 |
| Monte Sereno | - | - | - | - | 1,506 | 2,847 | 3,434 |
| Morgan Hill | 600 | 900 | 1,000 | 1,600 | 3,151 | 5,579 | 17,060 |
| Mountain View | 1,000 | 3,300 | 3,900 | 6,600 | 30,889 | 54,132 | 58,655 |
| Palo Alto | 5,900 | 13,700 | 16,800 | 25,500 | 52,287 | 56,040 | 55,225 |
| San Jose | 39,700 | 57,700 | 68,500 | 95,300 | 204,196 | 459,913 | 628,283 |
| Santa Clara | 5,200 | 6,300 | 6,700 | 11,700 | 58,880 | 86,118 | 87,746 |
| Saratoga | - | - | - | - | 14,861 | 26,810 | 29,261 |
| Sunnyvale | 1,700 | 3,100 | 4,400 | 9,800 | 52,898 | 95,976 | 106,618 |
| Total Incorporated | 59,800 | 92,100 | 109,200 | 161,000 | 481,433 | 922,898 | 1,167,013 |
| Total Unincorporated | 40,900 | 53,000 | 65,000 | 129,500 | 160,882 | 142,415 | 128,058 |
| County Total | 100,700 | 145,100 | 174,900 | 290,500 | 642,315 | 1,065,313 | 1,295,071 |

a. Annexed into the city of San Jose

Note: Dash indicates city not yet incorporated

Sources: Santa Clara County Planning Department, Housing Characteristics, Cities, Santa Clara County, 1970 (San Jose, Calif.: Santa Clara County Planning Department, 1971); and Santa Clara County Planning Department, Advanced Final Count of 1980 (San Jose, Calif.: Santa Clara County County Planning Department, April 1981).

university or who sought the genteel tranquility of gracious living within a railroad commute of the office towers of San Francisco. A more opposite ambience from that of San Jose and surroundings could not be imagined, yet their fates were already intertwined.

By the 1930s, Frederick Terman, a visionary electrical engineering professor at Stanford, had perceived the lucrative partnership that could be shared by industry and the academy, and he encouraged his talented students to establish their fledgling enterprises nearby rather than pursue employment in the East. Thus, he convinced David Packard to return to Palo Alto from Schenectady, New York, and General Electric. There he joined William Hewlett, another Terman protegé, and, in a small garage shop, the Hewlett-Packard Company began operations in 1939; it now employs about 30,000 worldwide, 40 percent of whom are in Santa Clara County. About the same time, Sigurd and Russell Varian, with $100 for supplies from Stanford and free use of its laboratories, developed the radar-essential klystron tube. The university's royalties on this one item have amounted to more than $2 million over the last 30 years. Today the multitudinous spinoffs of the Stanford atelier are staggering in their proliferation. For example, an offshoot of William Shockley's pioneering transistor firm became Fairchild Semiconductor, which has in turn spawned at least 38 other enterprises. Similarly, Varian alumni formed Spectra-Physics, now the world's largest laser company.[33]

Growth continued to mount. The personnel imported by war material industries during the early 1940s and returning military contingents from the Pacific Theater found that the San Francisco Bay area could absorb them into its expanding labor force. Indeed, many of the new jobs were located in the northern portion of Santa Clara County, particularly in the swath from Palo Alto southeast to Sunnyvale and Santa Clara. San Jose interests had recognized the enormous potentials for growth heralded by the postwar era and had mobilized aggressively to attract its share. In 1944, local politicians and businessmen formed the Progress Committee in order to achieve that mandate. Landowners, the construction trades, retailers, bankers, realtors, and other mercantile groups managed to elect the first in a long succession of city councils with an enthusiastic commitment to growth. In the 1950s, when questions were raised concerning the impact of this policy and the concentration of wealth and influence perceived to be its prime mover, a similar axis emerged with the thinly veiled title of the 'Book of the Month Club'. Although it was averred that the main purpose of this clique was to discuss certain literary works, its membership of businessmen, builders, realtors, land speculators, the publisher of the San Jose Mercury and News (the city's only morning and afternoon papers), San Jose's city manager, councilmen, and planning commissioners, among others, signalled an interest in matters far less altruistic.[34] Indeed, they had much to occupy them, for the county's population had grown to 290,000 in 1950, an increase of two-thirds since 1940. The lush farmlands on the valley floor were being sold for development, and attendant residential services followed in rapid order. Unable to garner the industrial base which tended to cluster around Palo Alto area agglomerations, San Jose cast its lot as the bedroom community for individuals working in the north county technology belt. Perceiving this aggressive posture on the part of San Jose, officials of other communities moved to secure their share of the prosperity.

Perhaps the key reason for the ensuing scramble in Santa Clara County's helter-skelter market in the early 1950s was the aggressive program of annexation pursued by San Jose. Between 1950 and 1980, it engulfed 140 square miles, thus increasing its total area to 157. Not an insignificant portion of this turf was brought into the city by means of extending fingers of city land to outlying tracts deemed ripe for development. One of these corridors was no less than three miles in length and a mere roadway in width.[35] This situation was viewed with alarm not only by proponents of rational growth but by other local interests as well. As a result, existing municipalities, such as Santa Clara and Sunnyvale, moved to consolidate their geopolitical holdings, while seven other localities experiencing sufficient nascent awareness, tempered by a reluctance to become swallowed up by San Jose, moved to implement a policy of incorporation. Thus, Campbell, Milpitas, and Cupertino, adjacent localities directly threatened by the expansion of San Jose, carved out their spheres of influence. It had been 40 years since the last incorporation had occurred in the county (see Figure 2.2 and Table 2.4).

At the same time, the eruption of hostilities in Korea had acted to provide additional momentum to

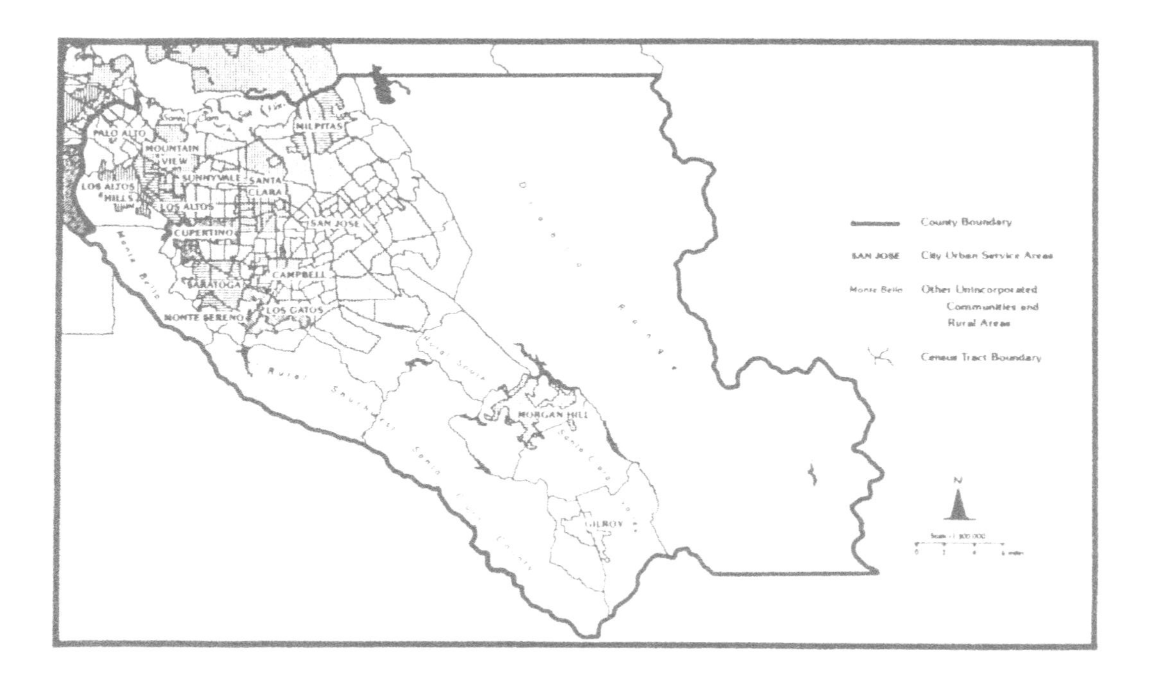


FIGURE 2.2 THE SAN JOSE METROPOLITAN AREA (SANTA CLARA COUNTY)

Source: Association of Bay Area Governments, Projections 79 (Berkeley: ABAG, 1979), p.IV-29

TABLE 2.4 DATES OF INCORPORATION FOR SANTA CLARA COUNTY MUNICIPALITIES

| Community | Year of Incorporation |
|---|---|
| San Jose | 1850 |
| Santa Clara | 1852 |
| Gilroy | 1870 |
| Los Gatos | 1887 |
| Palo Alto | 1894 |
| Mountain View | 1902 |
| Morgan Hill | 1906 |
| Sunnyvale | 1912 |
| Campbell | 1952 |
| Los Altos | 1952 |
| Milpitas | 1954 |
| Cupertino | 1955 |
| Los Altos Hills | 1956 |
| Saratoga | 1956 |
| Monte Sereno | 1957 |

Source: Santa Clara County Planning Department (1978)

the vigorous industrialization that was proceeding in the county. Sylvania, Fairchild, FMC, Admiral, Kaiser, General Precision, and Lockheed made their move into the area, and San Jose managed to attract facilities that included Ford, General Electric, and IBM, among others. This industrial pattern, consolidated in the post-World War II years, continued to develop in the 1950s and beyond and provided the structural underpinnings for future developments. The entrepreneurial and scientific vigor of local researchers spawned the aerospace industry and then a burgeoning of electronics and other related high technology enterprises. Mergers and acquisitions further increased the number of significant operations.

The ultimate in Terman's industry-academy tandem is represented by the Stanford Industrial Park. Because its founder prohibited the sale of the land in his bequest, Stanford developed 660 acres adjoining the campus for industry. With 55 tenants, including Hewlett-Packard and Varian, the park employs over 17,000 persons and has been an important factor in the continuing creation of a swarm of smaller firms, many of which have grown into major corporations which in turn have spun off other enterprises. By 1977, Santa Clara County was the established national center of electronics and high technology, with nearly 200,000 employed in that sector alone.[36]

Population growth was a simultaneous response to the rapid industrialization that was occurring. While each new manufacturing job generated about 1.2 additional positions in the matured economy of the San Francisco Bay area in 1975, this multiplier was two to three times higher in the early 1950s when Santa Clara County was beginning to urbanize.[37] Other cities in the county, as noted above, each reacted to the aggressive expansion of San Jose. For example, Santa Clara and Sunnyvale, two of the most proximate, consolidated their territorial spheres of influence and effectively capitalized on their convenient access to the Palo Alto technology orbit by attracting a large industrial base as well as encouraging substantial residential construction at reasonable cost. It is hardly surprising that these two communities experienced the highest rates of growth in the frenetic pace of urbanization that inundated the county during the 1950s. At the same time, Cupertino incorporated so that its destiny could be better controlled than if San Jose were allowed to assume jurisdiction. The legacy of this action is obvious. By pre-empting San Jose, Cupertino could achieve a more balanced tax base by eschewing dependence on a broad-based residential growth and create a more exclusive housing stock by allowing the market to establish more costly land prices.

During the 1950s, 1960s, and 1970s, the development of housing proceeded apace as all conditions were favorable to the development of large tracts of single-family housing. The valley floor was easily urbanized and highly accessible once the circulation system developed and matured. Financing was readily available, and the large supply of land kept housing prices low, even though the demand generated by immigration was vigorous. The single-family house was the preferred mode of living under these bountiful circumstances, and it was a goal attainable by nearly three-quarters of the population by 1970.[38]

By the early 1970s, however, there were signs that the supply of developable land was taking on finite proportions. As land costs began to escalate, lots became smaller, and townhouse and condominium projects emerged. Further, environmental and fiscally inspired moratoria put a brake on housing supply, thus fueling the flames of inflation in the market. The supply of large parcels began to dry up, and 'in-filling' developments of less than 50 units started to appear. The northern half of Santa Clara County is now a mature, fully urbanized community with most of its growth already accomplished. Redevelopment to higher densities in some areas is now the only alternative posed by the near future.

One can perceive the increments of growth of each period in the two centuries that the county has been inhabited by other than its indigenous population of the California Indian group, collectively known as Costanoan. There are old nineteenth-century kernels in Campbell, Los Gatos, downtown San Jose, and so forth. There is the development that accompanied the agrarian and food-processing centers before World War II. The suburban sprawl that consumed most of the valley after World War II is the dominant feature of the landscape today. Finally, here and there are the in-fillings of the mid and late 1970s, small projects, mostly of higher density. This last group will probably evolve into the type of unit that will be defined by economic constraints and by the lifestyle preferences of future populations.

## B. ISRAEL: IMMIGRATION AND DEVELOPMENT IN THE PROMISED LAND

Although Jews were present in Palestine continuously for millenia, their distinctive settlements emerged only in the last quarter of the nineteenth century. Eclectic in origin, they were integrated into an existing system of Arab agrarian villages and of towns which dated from biblical times.[39] Examples of the latter are the coastal cities of Acre and Jaffa which have been historically important trade and cultural centers, and Jerusalem, situated in the mountains to the east. Until security conditions improved in the years preceding the First World War, Arab settlements were confined to the rugged terrain east of the central plain. Thereafter, these indigenous villages penetrated westward into the coastal plain bordering the Mediterranean.[40]

A substantial part of Jewish population growth in Palesine was due to a series of waves of immigration (Aliya), primarily from central and eastern European countries. Communities were generally rural settlements based on the immigrants' Zionist ideology which stressed rebuilding the Jewish national homeland through agricultural development.[41] Often built on undesirable sites in the coastal plain which were purchased from Arab landowners, these Jewish rural settlements may be grouped into four basic forms:

> Moshavot - European-type villages, with land, buildings and equipment in individual ownership;
> Kibbutzim - Highly ideological communal agricultural settlements with virtually no private property;
> Moshavim - Cooperative settlements with private households and labor but collective regulation of land ownership, equipment, and marketing;
> Others - Settlements with some characteristics of the above, such as the Moshav Shittufi, a collective of small farm villages with communal production and private family life.[42]

With continued Jewish immigration (approximately 500,000 arrived in the six decades prior to the end of World War II) these rural settlements were established in progressively more remote locations away from the coastal area to the hills, foothills, and desert areas (see Figure 2.3). As Table 2.5 shows, the first Aliya began in 1882, with six others following up to independence in 1948.

As a result of World War I, Palestine was transferred from Turkish Ottoman to British Mandatory administration. Despite British opposition to Jewish immigration, several important political and ideological organizations emerged in the housing market for the purpose of assisting the growing numbers of immigrants to find adequate shelter. During the 1920s and 1930s, these organizations, usually cooperative housing societies associated with the General Federation of Labor (Histadrut), political parties, or religious groups, built low- and moderate-cost housing on relatively inexpensive land outside of the major urban centers.[43] For example, during the 1928-44 period, one of these societies, the Shikan Workers Company Limited, built 6,600 homes in such low-cost garden suburbs as Kiryat Hayim near Haifa, and Kirayat Avoda outside of Tel-Aviv.[44]

Meanwhile, Tel-Aviv and Haifa were also growing rapidly. This was due to the concentration of labor and resources required in the early phases of industrialization and to the urban preferences of middle class immigrants of European professional and business backgrounds.[45] For example, the city of

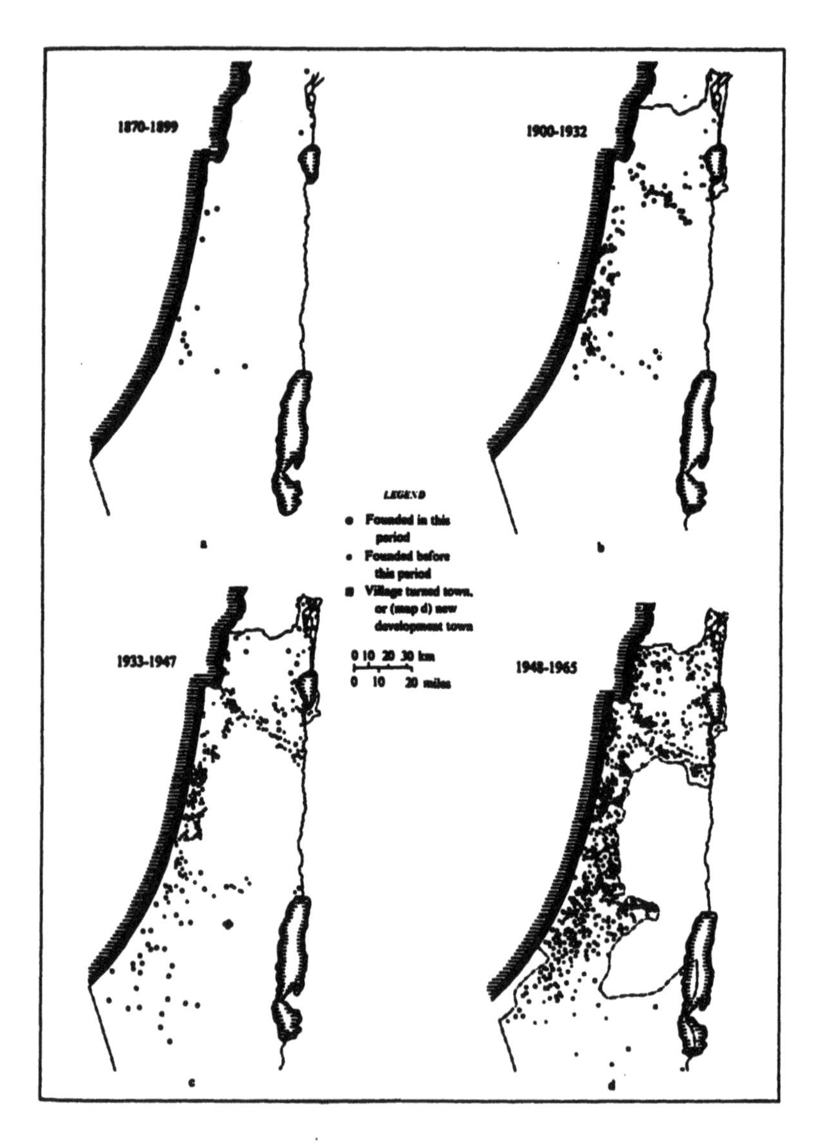


FIGURE 2.3 GROWTH OF JEWISH RURAL SETTLEMENTS IN PALESTINE/ISRAEL: 1870-1965

Source: Efraim Orni and Elisha Efrat, <u>Geography of Israel</u> (Israel University Press, Jerusalem, 1980), p.241

TABLE 2.5 IMMIGRATION TO PALESTINE/ISRAEL: 1882-1981

| Migration | Size (thousands) |
|---|---|
| First Aliya (1882-1903) | 20-30 |
| Second Aliya (1904-1914) | 35-40 |
| Third Aliya (1919-1923) | 35 |
| Fourth Aliya (1924-1931) | 82 |
| Fifty Aliya (1932-1938) | 217 |
| World War II (1939-1945) | 92 |
| Post World War II (1946-May 1948) | 61 |
| Statehood (1948-1951) | 684 |
| (1952-1954) | 51 |
| (1955-1957) | 161 |
| (1958-1960) | 72 |
| (1961-1964) | 220 |
| (1965-1968) | 72 |
| (1969-1972) | 172 |
| (1973-1981) | 244 |

Sources: 1882-1948 data from Moshe Sicron, Immigration to Israel (Falk Project and CBS, Jerusalem, 1957), p.21; 1948-1981 data from Elisha Efrat, Urbanization in Israel (Croom Helm, London, 1984), p.22

Tel-Aviv, which was founded in 1909 was a suburb of the ancient coastal city Jaffa, ironically grew to become the nexus of numerous satellite communities by the 1940s, and eventually Israel's largest metropolitan area.[46] As the major coastal gateway to Palestine, Tel-Aviv became the Mandate's most important economic center for industry, commerce, banking, and insurance as well as a prominent focus for cultural, political, and social activities. The housing industry was not able to keep pace with the unprecedented growth; by 1944, one half of Tel-Aviv's 167,000 residents lived in housing units with three or more persons per room.[47] At the time of independence in 1948, the Tel-Aviv area and the surrounding Central District in the coastal plain contained over half of the national population.[48]

1. Independence and Suburbia

While the pace of housing construction quickened immediately after World War II, it was not until the late 1940s that a massive housing program unfolded. The new Israeli government quickly passed the Law of Return, which guaranteed every Jew the right to settlement, and to Israeli citizenship. At first, the young nation was overwhelmed with the unprecendented avalanche of migration from all over the world which doubled the Jewish population in the first three years of statehood. Many of these new arrivals were homeless refugees with little or no resources. Consequently, the State had to play a major role in the housing field and undertook emergency means to provide shelter for the soaring population which was arriving at a rate of more than 20,000 persons per month.[49]

Initially, housing units were situated in former British Army camps and in Arab communities abandoned during the 1948-49 Arab-Israeli war.[50] When these quarters were exhausted, it became necessary to establish transit settlement camps, Maabarot, in the vicinity of existing towns and areas of development. The next stage involved the transfer of the immigrants to semi-permanent homes.[51] During the 1950s, when government policy channeled low and moderate-income immigrants to the outlying districts of the major urban areas, a settlement pattern arose that was more typical of Third World than of industrialized countries. Thus, remote sites were developed in order to keep housing costs down.[52]

Despite familiarity with the British garden city tradition, during the 1950s Israel built many new suburban housing units in environments of very uneven quality.[53] This was due primarily to the tremendous pressures to build expeditiously and to the limited capacity and ability of a fledgling construction industry.[54] In addition, the social and cultural diversity of the immigrants and the inaccessibility of new residential development added to the difficulties of these early suburbs.[55] Nevertheless, everyone in the nation was provided with shelter and both the quantity and quality of housing improved dramatically over time.[56] As Table 2.6 shows, by the mid-1950s, the number of housing units built annually per thousand population in Israel was more than double that of most European nations, themselves rebuilding vigorously after the destruction of World War II; and as Table 2.7 indicates, during the 1960s and 1970s the number of persons per room declined substantially.

Another ingredient in the Israeli suburban mix is an array of rural settlements which gradually

TABLE 2.6 THE SCOPE OF BUILDING IN ISRAEL AND SEVERAL INDUSTRIALIZED COUNTRIES: 1950-56

| Country | Number of Housing Units built annually in the country concerned (thousands) | | | | | | | No. of Units built per 1,000 pop. in 1956 | Size of the population in 1956 (thousands) |
|---|---|---|---|---|---|---|---|---|---|
| | 1950 | 1951 | 1952 | 1953 | 1954 | 1955 | 1956 | | |
| Austria | 46.2 | 13.0 | 57.3 | 38.2 | 40.5 | 41.6 | 42.0 | 6.0 | 6,983 |
| Belgium | 44.7 | 35.5 | 33.3 | 39.2 | 44.9 | 44.6 | 42.8 | 4.8 | 8,925 |
| Czechoslovakia | 38.2 | 30.9 | 39.3 | 39.0 | 38.2 | 50.6 | 63.7 | 4.8 | 13,228 |
| Denmark | 20.4 | 21.5 | 19.0 | 21.3 | 23.3 | 24.0 | 19.8 | 4.4 | 4,466 |
| Finland | 26.0 | 28.5 | 31.2 | 28.9 | 31.0 | 33.2 | 31.9 | 7.4 | 4,291 |
| France | 70.6 | 76.7 | 83.9 | 115.5 | 162.0 | 210.1 | 236.5 | 5.4 | 43,648 |
| Greece | 48.7 | 43.9 | 59.9 | 51.5 | 46.3 | 53.3 | 55.4 | 6.9 | 8,031 |
| Netherlands | 54.8 | 64.8 | 57.4 | 62.6 | 70.5 | 61.9 | 69.2 | 6.4 | 10,888 |
| Norway | 22.4 | 20.9 | 32.7 | 35.1 | 35.4 | 32.1 | 27.1 | 7.9 | 3,462 |
| Switzerland | 25.0 | 30.0 | 27.5 | 28.4 | 36.1 | 39.4 | 39.4 | 7.8 | 5,039 |
| Soviet Union | - | - | - | 1245.0 | 1351.0 | 1512.0 | 1613.0 | 8.0 | 201,060 |
| Great Britain | 214.7 | 209.4 | 354.5 | 330.4 | 350.7 | 328.6 | 310.0 | 6.1 | 51,208 |
| Israel* | 22.2 | 36.4 | 30.1 | 25.6 | 23.0 | 27.2 | 28.2 | 15.1 | 1,872 |

* These figures do not include temporary and semi-permanent housing.

Source: Haim Darin-Drabkin, 'Economic and Social Aspects of Israeli Housing' in Haim Darin-Drabkin (ed.), Public Housing in Israel: Surveys and Evaluations of Activities in Israel's First Decade (1948-1958) (Gadish Books, Tel-Aviv, 1959), p.42

TABLE 2.7 ISRAELI HOUSEHOLDS BY PERSONS PER ROOM: 1959-77

| Year | Percent Families with Two or More Persons per Room[b.] | Percent Families with Three or More Persons per Room[a.b.] |
|---|---|---|
| 1959 | 56.9 | 22.9 |
| 1969 | 27.7 | 8.5 |
| 1974 | 20.4 | 5.2 |
| 1977 | 15.3 | 2.9 |

a. By 1982, this criterion was obsolete.
b. Lavatories and kitchens do not count as rooms. If additional criteria are applied, such as the need for repairs or private family baths or showers, the number of families in substandard dwellings could be 45,000 to 50,000 (1977).

Source: Daniel Shimshoni, Israel Democracy: The Middle of the Journey (The Free Press, New York, 1982), p.292

became absorbed into the expanding rings of major metropolitan areas. First, there were some thirty Moshavot, agricultural settlements based on private ownership of land and private enterprise.[57] Located primarily near the large urban centers in the coastal plain, many of these were transformed fully or partially into middle class suburban neighborhoods and thirteen have even grown large enough (at least 10,000 people) to receive municipal status, such as Herzliyya, or Kefar Sava in the Tel-Aviv region, an area surveyed in the study detailed in the chapters which follow.[58] Second are the numerous collective agricultural communities (Kibbutzim) which, although preserving their rural character, have become engulfed by (and part of) the expanding suburban ring.[59] Finally, many rural Arab villages have been engulfed by suburban growth, often in the capacity of low- and moderate-income areas due to their high residential densities, poorer quality housing, narrow streets, and inadequate infrastructure.[60]

As in other industrialized nations with highly developed urban cores, Israeli metropolitan decentralization is superimposing a growing number of middle-class families as well as commercial and

TABLE 2.8 POPULATION IN THE TEL-AVIV METROPOLITAN AREA: 1948-1981

| Area | 1948 | | 1961 | | 1967 | | 1981 | |
|---|---|---|---|---|---|---|---|---|
| | No. (thousands) | % Total | No. (thousands) | % Total | No. (thousands) | % Total | No. (thousands) | % Total |
| Central City | 248 | 83 | 386 | 60 | 388 | 51 | 358 | 42 |
| Suburban Ring | 51 | 17 | 264 | 40 | 368 | 49 | 495 | 58 |
| Total | 298 | 100 | 650 | 100 | 756 | 100 | 853 | 100 |

Source: Elisha Efrat, *Urbanization in Israel* (Croom Helm, London, 1984), p.69

industrial activities on an already complex and socially heterogeneous pattern of suburbanization.[61] For example, during the 1948-81 period, the Tel-Aviv suburban ring grew from about 50,600 to 495,000 persons and increased its share from 17 to 58 percent of the total population of the Tel-Aviv metropolitan area, while the central city actually lost population after 1967 (see Table 2.8). A similar pattern of suburban growth around other major cities has also been documented with the technical exception of Jerusalem, which incorporated most of its metropolitan area within its municipal boundaries after reunification in 1967.[62] Within the boundaries of Jerusalem, decentralization is exemplified by large-scale developments in areas such as Ramat Eshkol to the north, and at Gilo to the south; the latter is also a study area in the following survey.[63] Thus, with all their variations, Israeli suburbs have become a mosaic dominated by increasingly middle-income residents.[64] This pattern has evolved as a result of (a) the rising social status of immigrants; (b) physical improvements in the older neighborhoods; and (c) recent migration by younger, middle income families seeking affordable housing and accessibility to open space.[65]

2. National Decentralization

Not all new arrivals in Israel were settled in large urban centers or in peripheral zones of metropolitan areas. While it was clear that a more urban approach would be needed to accommodate large-scale immigration than the modest-sized agricultural communities of the pre-1948 era, it was also clear that a comprehensive policy of settlement would be needed on the national scale. As Elisha Efrat points out, the primary reasons for establishing such a national settlement policy were:

1. The mass immigration that occurred in the years immediately following statehood created a pressing housing shortage which only urban settlement could solve;
2. the desire to create a series of dispersed medium-sized towns as intermediate centers between the large cities and small agricultural villages; and
3. the need to populate frontier areas as a means of securing the State's territory.[66]

Despite these imperatives, agricultural settlements continued as a means of population dispersal because of the new nation's socialist-Zionist pioneering ideology and because of the strong political position of agricultural interests in the ruling Labor party.[67] Indeed, more than 280 rural settlements were established between 1948 and 1951. Further, the 1951 national plan targeted 22.6 percent of the total population to live in rural settlements.[68] Yet, the Jewish rural population, whose maximum of 370,000 was attained in the mid 1950s, declined to 265,000 by 1973.[69] This was due to limited availability of adequate soil and water resources and the unwillingness of many immigrants to live in rural settlements.[70]

Because of these limitations the building of new towns became the major thrust of population dispersal policy. Their purpose was to divert population growth away from the urbanized coast to outlying regions of the country for the purpose of balanced regional development, defense, and the absorption of new immigrants into the national social system. New towns were additionally important in that they created a regional structure that would integrate scattered populations into a size-graduated system of urban and semi-urban centers.[71] Based on geographic central place theories, the new settlement hierarchy was to be constructed in five levels, starting with small agricultural villages and building up to large towns, as follows:[72]

- Type A - Villages with 500 inhabitants
- Type B - Village centers with 2,000 inhabitants each.
- Type C - Semi-urban centers with 6,000 - 12,000 inhabitants each.
- Type D - 40,000 - 60,000 inhabitants each.
- Type E - Large towns with 100,000 inhabitants or more.[73]

Thus, in a given geographic region, the smallest units (Type A), consisting of miniscule rural settlements such as _Kibbutzim_, _Moshavim_, and small villages, would be accommodated by the next larger B units, rural service centers, which in turn would be serviced by a C center which would extend its functions to about 6,000 to 12,000 persons within a radius of 3 to 7 miles.[74] Level D towns were to be major centers for planning regions, each containing two to three C centers and a target population of about 60,000. Finally, Type E centers were to be the central places of major national regions.

In their effort to fill in the missing links of the urban hierarchy, Israeli planners initially concentrated on level C and D centers, a strategy which met with mixed success.[75] Despite government incentives to decentralize industry to these new developments, they were often too small to offer a rich enough variety of services to compete with existing metropolitan centers, usually a short travel time away; and frequently the new residents were opposed as non-pioneering urban immigrants by the veteran agricultural communities they were designed to serve.[76] Nevertheless, these centers accommodated about 17.5 percent of the one million immigrants during the first decade of statehood.[77]

By the 1960s, the policy had shifted to establish regional centers (D and E towns) which were more successful than the smaller developments due to their larger minimum population size of about 20,000, a critical mass for self-sustained economic growth and for the regional provision of a wide range of urban services.[78] These towns were also better able to accommodate, and benefit from, large-scale industrial decentralization from the metropolitan areas of Tel-Aviv and Haifa.[79] As with those of Great Britain, the larger Israeli new towns were designed to have several distinct neighborhoods and related services which helped create an improved quality of residential social life.[80] The largest of these, Beer Sheva, with a 1977 population of 101,000, became the regional center of the south and one of the most important cities in the nation.[81] By the mid 1970s there were over thirty new towns with a combined population of over 500,000, representing over 18 percent of the Jewish population in Israel (see Figure 2.4).[82]

No doubt substantial social adjustments were required by residents of these new towns, especially in the smaller and more remote settlements.[83] Still, many of the social and population dispersal objectives were at least partially achieved. Studies have shown that progress has been made in the social integration of various Jewish cultural groups in many new towns.[84] It has also been shown that the Tel-Aviv metropolitan area's share of the national population declined from 34 to 24 percent during the 1948-81 period,[85] while outlying regions, particularly the Southern District, increased their proportion.[86] In addition, the new town policy did help fill in the hierarchy of urban central places with many intermediate size towns.[87] Figure 2.5

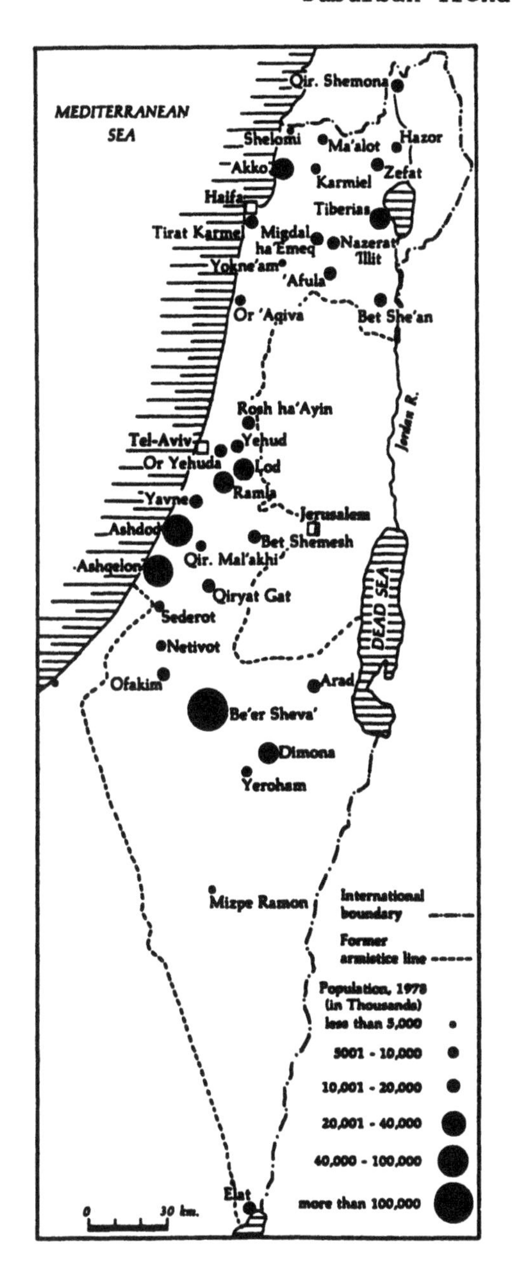


FIGURE 2.4 NEW TOWNS IN ISRAEL BY POPULATION SIZE: 1979

Source: Daniel Shimshoni, <u>Israeli Democracy: The Middle of the Journey</u> (The Free Press, New York, 1982), p.114

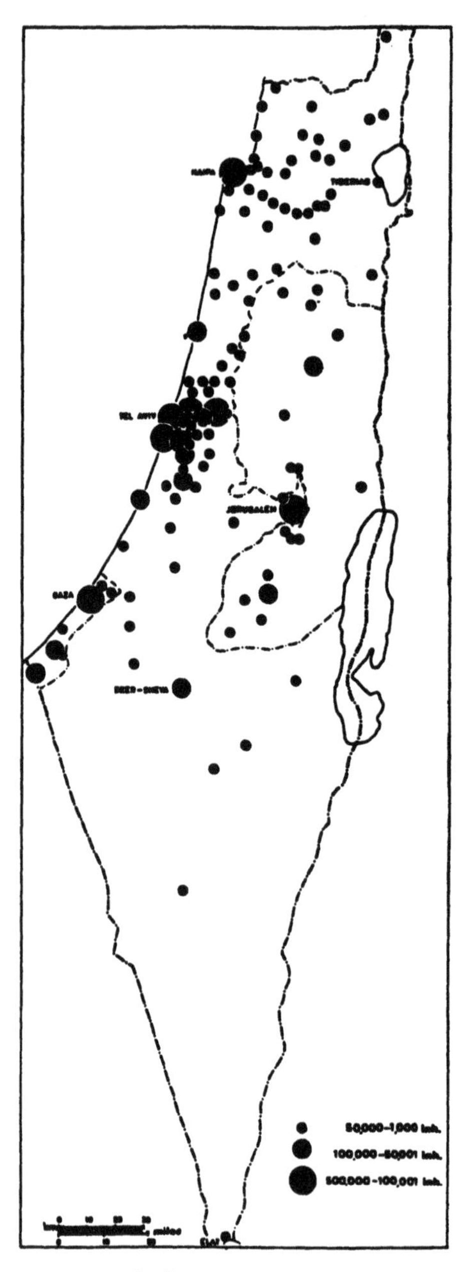


FIGURE 2.5 THE URBAN SYSTEM IN ISRAEL: 1983

Source: Elisha Efrat, Urbanization in Israel (Croom Helm, London, 1984), p.213

presents the resulting urban system in Israel.

To be sure, other forces, such as congestion and land costs, have encouraged market-induced decentralization within and between regions. Indeed, the well-documented congestion of urban core areas such as downtown Tel-Aviv,[88] and the recent growth and dominance of private sector housing units attest to the substantial influence of market forces on the decentralization process.[89] But it is also clear that Israeli public policies did foster much of the intra-urban and inter-regional decentralization of population and economic activity.[90] In doing so, Israel has helped to accelerate the social integration of its diverse population and its movement along the path of advanced industrial and urban evolution.

## C. THE NETHERLANDS: REGIONAL POLICY INITIATIVES, REBUILDING AND POST-WAR AFFLUENCE

The two decades which encompassed the successive convulsions of the Great Depression and World War II set the stage for profound changes in the industrialized world. While the postwar re-alignment of global antagonisms has captured its full share of headlines, much less is known of the internal problems that challenged the ingenuity of western nations recovering from the traumas of the 1930s and 1940s. As discussed in Chapter 1, the United States and Canada, unique in their geographical remoteness from hostilities in the European and Asian theaters, rushed headlong into an unprecedented era of prosperity fueled by relentless suburbanization, mass automobile ownership, and a consumer-based economy spurred on by two decades of exuberant natural population increase.

However, most European nations encountered problems far beyond the scope of those with which the architects of North American postwar prosperity had to cope: entire economies had to be rebuilt; acute housing shortages demanded a swift response; vigorous population growth required major adjustments in social, spatial, and economic planning; and the question of decolonization, with its explosive emotional, political, and economic ramifications, also thrust itself into the embroilment. Few nations faced greater pressures from these challenges than did the Netherlands. In this section, then, the Dutch response to these issues will be discussed with specific emphasis on (1) regional policy and urbanization, and (2) the determinants of postwar housing policy.

## 1. Regional Policy, Economic Planning and Urbanization in the Netherlands since 1945

The prosperity and economic growth enjoyed by the Dutch in recent decades is inextricably linked with that experienced by other western European nations. However, despite the resurgence of neighboring market-based economies, the recovery of the Netherlands was by no means assured. Because it had been slow to follow the model of British industrialization, Dutch manufacturing was particularly vulnerable in the first half of the twentieth century. Not only were natural resources in short supply, but the labor force lacked a high level of training; further, capital was scarce due to the entrepreneurial investments on which the Dutch had relied since the seventeenth century. Although the necessary improvements in infrastructure, internal waterways and railroads had been completed by the 1870s, advances favored light industry rather than the production of capital goods.[91] Even though the Netherlands maintained neutrality during World War I, the conflict stymied its industrial modernization and living standards suffered. These conditions persisted during the 1920s and were further aggravated by the economic instability of that decade. Consequently, the Dutch were more ill-prepared than were their neighbors for the Great Depression, an episode from which the Netherlands was slower to recover than any other country in Western Europe.[92]

Economic activity continued to decline during the German occupation. Further, damages from the war were catastrophic. Transportation and communications, vital to the nation's well-being, lost 55 percent of their productive capacity; the merchant marine's capabilities were cut in half; and of 300 electric railroad engines, only five were in use by 1945. Further, 40 percent of manufacturing capacity had been destroyed, and at least 10 percent of prime agricultural lands had been inundated.[93] The agenda which the Dutch faced in 1945 was monumental. The productive base in all branches of industry had to be rebuilt and the nation's finances restored to a position of international equilibrium. Because of the Netherlands' small domestic market and need for imported raw materials, manufacturing had to be redirected to the most modern sectors that would best lend themselves to both export abroad and minimizing the need for imported goods.

Consequently, chemicals, metallurgy, electrotechnical goods, and energy works provided the foundation for the remarkable economic growth of the postwar era.[94]

A second challenge centered on the creation of a policy which would nurture national economic recovery while at the same time achieving an equitable distribution of the costs required by this rigorous mobilization of resources and capital. This course of action was enunciated by the Minister of Economic Affairs in the 1949 Memorandum on the Industrialization of the Netherlands. As a general economic policy, industrialization could only be achieved with full employment, stable prices, and a favorable balance of payments. It was realized that increases in productivity should not be channeled to individual consumption, but must be re-invested to further the continued rebuilding of national industrial capacity. 'It is, in effect', the Memorandum concluded, 'a question of austere living and of exerting our energies to avoid greater adversity in the future'.[95] Crucial to this equation were wage restraints, controlled rents in a period of an acute housing shortage, and infusions of public subsidies to achieve production of new housing units.[96]

A third problem which arose in the post-1945 period was the loss of the Indonesian archipelago. Not only did this represent the removal of a huge market for exports (9 percent of Dutch output in 1939) but also it was expected that by 1949 decolonization would erode the nation's income by some 10-15 percent.[97] In addition, an army of more than 100,000 was maintained in Indonesia at great cost which failed to achieve results that were in the interest of either nation. Concern about Indonesia was reflected in the realization that, by 1939, one-sixth of Holland's wealth had been invested in the East Indies.[98] Given the demands for the reconstruction of the economy of the Netherlands after the war, the huge capital losses suffered by Dutch investors and exporters were a necessary price to pay for the dispossession of Indonesia. The colony would have jeopardized the national recovery, for it, too, required massive re-investment in the wake of the ravages wrought by the Japanese occupation.[99]

The Netherlands, then, ranks as one of the major economic successes of postwar Europe despite the exactions of war at home and abroad. Its industrious labor force and consensus-forming political institutions paved the way for the nation's return to prosperity. However, while it was recognized that prosperity depended upon a favorable balance of international accounts, the way in which this was achieved must relate to internal economic activities and their spatial distribution. Indeed, by the 1950s it was realized that economic growth, full employment, and a desirable income distribution were factors directly influenced by physical and socio-economic planning policy.[100]

In considering the spatial structure of the Dutch economy, there were striking regional characteristics with which the planners of postwar recovery had to reckon. In the northern provinces of Drenthe, Friesland and Groningen, agriculture was dominant (see Figure 2.6). But as this sector became less labor-intensive, unemployment soared, reaching 11.3 percent by 1952 in areas near the German border.[101] Prospects in the southern provinces of Limburg and North Brabant seemed to be more favorable, but poor transportation links, a weak industrial tradition, and the high birth rate of the largely Roman Catholic population contributed to an unemployment rate second to only the North.[102] In contrast, the heavily urbanized West of the Netherlands, which contains the major cities of Amsterdam, The Hague, Rotterdam and Utrecht, possessed not only considerable concentrations of manufacturing, but also world-class port facilities, exemplified by those of Rotterdam. Further, a growing and vital services sector characterized what van de Knaap terms 'the city-rich West'.[103]

Therefore, if Dutch postwar economic objectives were to be attained, regional imbalances would have to be addressed. In the early 1950s, the first legislation was passed which provided aid for areas of high unemployment, particularly in the northeastern province of Drenthe. This was accomplished with subsidies for individual migrants as well as for enterprises eligible for investment incentives. In addition, the government allocated funds for industrial estates, transportation improvements, and for the expansion of educational facilities. This approach was not uniquely Dutch, but, rather, marked the beginning of a growth center policy, a phenomenon found in other problem areas in Western Europe.[104] Such a strategy marked only the beginning of a period of refinement of regional economic planning in the Netherlands that is still in evolution.

Another problem was posed by the concentrations

FIGURE 2.6 DEVELOPMENT AREAS IN THE NETHERLANDS: 1969

Source: Francois J. Gay, 'Benelux', in Hugh D. Clout (ed.), Regional Development in Western Europe (John Wiley, New York, 1981), p.198

of economic activity in the western Netherlands. It was only in 1960 that the idea jelled that the horseshoe-shaped urbanization from Rotterdam through The Hague, north to Amsterdam and then east to Utrecht was in fact one large metropolitan area. However, by 1966, this view was modified to comprehend this 'Randstad', or rim city, as an organism with two wings, a northern sector beginning at Alkmaar, north of Amsterdam, and extending east through Utrecht to Arnhem, near the German border, and a southern wing commencing at Leiden running south to Rotterdam and then east as far as Gorinchem (see Figure 2.6).[105] Just as the North and Southeast of the Netherlands required assistance if an extended postwar national recovery were to be achieved, then so did the Randstad figure prominently in this strategy. Not only was it necessary to decentralize jobs and population to other regions, but, when in 1965, projections indicated that the Dutch population would increase by nearly 70 percent to 21 million by the year 2000, it became imperative to create counter-magnets outside the Randstad and to prevent urbanization of its open space, the so-called 'Green Heart'.[106]

The imperatives of open-space preservation and decentralization define planning policy in the Randstad. With a population density of over 2,200 per square mile in 1970, a figure twice the national average (itself the highest in Western Europe), demographic increases and smaller household sizes placed intense pressures on available land no matter how successful policies of decentralization might be. Further, while the importance of housing cannot be discounted, equally crucial is the thriving horticultural sector of the economy, most of which is contained within glasshouses requiring heavy capital investment per acre. In a cold climate where overcast skies predominate, the effusion of color in the form of fresh flowers makes the Westland glasshouse district near The Hague a necessity for nearly every Dutch household. More importantly, this area accounts for about 40 percent of the Netherlands' agricultural output, but the flat land which it occupies is well-suited for urbanization. In addition, the bulb fields between Leiden and Haarlem and the intensive garden cultivation in the same area rank as important contributors to national economic objectives and must be protected from urban intrusion.[107]

With the imperatives of open space preservation and decentralization of population and industry from the Randstad well-established by 1970, it became necessary to make some further adjustments to national policy. Not only did the expected rate of natural population increase significantly slacken, but efforts to channel industrial development and population to growth centers outside the Randstad fell short of expectations. With a downwardly revised population forecast of 15.5 million by the year 2000 (vs. the 21 million figure projected in 1965), greater emphasis has been placed on efficient growth increments within the Randstad while viewing the gradual population increase in the North as a satisfactory outgrowth of the decentralization policy of the 1950s.[108]

In accepting a more market-based urban policy for the Randstad, the Dutch government acknowledged that private economic and individual preferences cannot be wholly modified by official directive. Instead, they can be accommodated by concentrating new urban development along either an existing or new railway infrastructure. For example, the new town of Zoetermeer (target population, 120,000) about ten miles east of The Hague, was developed along a specially-built rail spur connecting it directly with the nation's administrative capital. Zoetermeer accommodates the need for additional residential development within the Randstad but simultaneously reflects government policy discouraging the growth of both intra- and inter-regional mobility. If efficient growth is not achieved, pressures will grow still more intense for urbanization of open space and of land with scenic qualities located within urban spheres of influence.[109]

However, despite the accomplishment of its regional initiatives, the Netherlands has not been more successful than other democratic societies in harnessing the whims of the marketplace; it should be emphasized that the Dutch government has been aware of the pragmatic limits of its authority. The Ministry of Housing and Physical Planning admits that 'policy-making takes place amidst a number of great uncertainties'.[110] Among these are international economic trends, developments in industry and technology, demography, and, indeed, the continued importance of the city itself.

While the Netherlands has not been immune to the process of suburban metropolitan decentralization, it has been able to exert some control over speculative land market and thus eliminate the fragmented, sprawling landscape that has determined the character of American metropolitan areas. A policy of advanced land acquisition by large municipalities and their virtual monopoly on the purchase of land needed for development has contributed to orderly growth and minimized leapfrog development. Nevertheless, decentralization has occurred for reasons which go beyond the policy of growth dispersal pursued since the 1950s. For one, large cities have lost population as a result of urban renewal activities, decreasing household size, desires for lower density housing in smaller population centers, as well as because of the disamenities associated with the older urban cores and the effects of ghettoization by recent immigrant groups. Suburbanization has also been abetted by the lack of coordination among the national government, provinces, and municipalities due to problems in arriving at a consensus concerning growth, activation of development schemes pre-dating agreements among the various levels of government, the absence of adequate growth management tools, as well as by pressures to increase total housing production by relaxing controls on the private sector.[111] Lastly, the material prosperity of the Netherlands has provided increasing sovereignty for the exercise of individual preferences, a point which has been stressed heavily by Berry in his examination of counter-urbanization in the United States.[112]

If metropolitan decentralization has been the general hallmark of settlement patterns throughout the industrial world, could a similar claim be made for the Netherlands in particular? Certainly, policies fostered by the Dutch government have de-emphasized the growth of the country's four major agglomerations (Amsterdam, Rotterdam, The Hague, and Utrecht) in favor of incentives to populate and industrialize the North, East and South. As discussed in Chapter 1, trends in Western Europe suggest that counter-urbanization has emerged in some nations as a significant phenomenon.[113]

In the Netherlands, data suggest that larger agglomerations are losing population and that smaller communities are experiencing increases. This decentralization occasioned by migration patterns is borne out by Table 2.9. In general, municipalities with populations in excess of 100,000 declined from 28.7 percent of the national population to 25.3 percent between 1971 and 1980 while rural municipalities and the urban countryside each modestly increased their shares of the national

TABLE 2.9 POPULATION OF MUNICIPALITIES IN THE NETHERLANDS: 1971-80

| Municipalities by Degree of Urbanization | February 28th 1971[a] | | January 1st 1975 | | January 1st 1980 | |
|---|---|---|---|---|---|---|
| | Number of municipalities | % of the total population | number of municipalities | % of the total population | number of municipalities | % of the total population |
| Rural municipalities | 269 | 11.1 | 259 | 11.4 | 239 | 11.6 |
| Urbanized countryside | 484 | 34.0 | 463 | 35.1 | 452 | 36.3 |
| Industrialised rural municipalities | 347 | 20.8 | 334 | 21.5 | 326 | 22.2 |
| Specific resident municipalities of commuters | 137 | 13.2 | 129 | 13.6 | 126 | 14.1 |
| Urban municipalities | 120 | 54.9 | 120 | 53.4 | 120 | 52.1 |
| large towns (100,000 inhabitants or more in built up area) | 14 | 28.7 | 15 | 27.2 | 15 | 25.3 |
| Total | 873 | 100.0 | 842 | 100.0 | 811 | 100.0 |

a. Results of the General Population Census.

Source: Statistical Yearbook of the Netherlands (Staatsuitgeverij, The Hague, 1980)

population.

When the four major cities of the Randstad are considered, the evidence suggesting decentralization is still more impressive. With the exception of the Utrecht agglomeration, each lost population from 1970 to 1979, while the Utrecht municipality registered the second greatest population decline (see Table 2.10). The fact that this latter agglomeration did experience a population increase reflects the urbanization pressures on the scenic forest and hill country to the east. For the Netherlands, this theater of potential conflict has arisen out of the uncertainties which surround the policy-making process. While regional imbalances pre-occupied physical and economic planners, the growth potential of small municipalities was overlooked.[114] According to data published in 1981, the Dutch expected municipalities of 50,000 or less to grow by 55.9 percent between 1960 and 1970; the actual growth amounted to 81 percent. Projections were far more accurate for cities of 50,000 - 100,000, 100,000 - 200,000 and 200,000 - 500,000. However, for the three municipalities in excess of 500,000 population, an anticipated 15.6 percent population increase for 1960-70 was in actuality a loss of 10.1 percent.[115] This decline has been complemented by the growth of smaller municipalities, a phenomoneon arising from affluence, improved communication and infrastructure, local aspirations, and the problem of controlling land prices and speculative development in some areas of the country.[116] In sum, these recent spatial and population growth patterns place the Netherlands among the more advanced urban-industrialized nations which are also experiencing a similar decentralization throughout their respective settlement systems.

This brief description of Dutch spatial and economic planning constitutes a summary of the components of a national policy which has attempted to coordinate actions designed to bring prosperity to a nation severely affected by the Great Depression and by the Second World War. Although the issues with which the Dutch are coping are more or less common to other Western European nations, it is notable that these policies represent a coordinated approach to economic, spatial, and social concerns. Such a convergence would be dismissed as the most radical form of utopian thought in the United States. While it may be on the agenda of some European nations with mixed economies, none has

TABLE 2.10 POPULATION CHANGE IN THE RANDSTAD, THE NETHERLANDS: 1970-79

| Agglomeration | 1970 | 1975 | 1978 | 1979 | % Change 1970-79 |
|---|---|---|---|---|---|
| Amsterdam | | | | | |
| Municipality of Amsterdam | 831,463 | 757,958 | 728,746 | 718,577 | -13.6 |
| Other municipalities | 208,932 | 232,832 | 236,500 | 239,154 | +12.6 |
| Total | 1,040,395 | 990,790 | 965,246 | 957,731 | - 7.9 |
| Rotterdam | | | | | |
| Municipality of Rotterdam | 686,586 | 620,867 | 590,312 | 582,396 | -15.2 |
| Other municipalities | 374,667 | 411,285 | 426,824 | 432,412 | +14.0 |
| Total | 1,061,253 | 1,032,152 | 1,017,136 | 1,014,808 | - 4.4 |
| The Hague | | | | | |
| Municipality of The Hague | 550,613 | 482,879 | 464,858 | 458,242 | -18.0 |
| Other municipalities | 168,813 | 196,026 | 208,533 | 213,239 | +26.3 |
| Total | 719,426 | 678,905 | 673,391 | 671,481 | - 6.7 |
| Utrecht | | | | | |
| Municipality of Utrecht | 278,966 | 256,016 | 240,713 | 236,053 | -15.4 |
| Other municipalities | 176,112 | 206,001 | 231,184 | 240,339 | +36.4 |
| Total | 455,078 | 462,017 | 471,897 | 476,392 | + 4.7 |

Source: Statistical Yearbook of the Netherlands (Staatsuitgeverij, The Hague, 1980)

made such bold or far-reaching attempts at national planning as the Dutch.[117]

## 2. The Determinants of Housing Policy in the Netherlands

Economic policy after the Second World War was the outgrowth of a long-lived national consensus predicated on austerity. If investment were to be successfully channeled to the productive sectors of the economy - those which would minimize the need for imports and maximize exports - then a 'social contract' would be required so that controlled wages would be linked to high employment and stable prices. Accordingly, the demands of the consumer sector and the accumulation of individual affluence were suppressed in favor of a policy of collective consumption and productive investment.

Because the Netherlands is heavily dependent on foreign trade for its economic well-being, postwar recovery policies required flexibility and, above all, that costs of production be competitive with those of nations with which the Dutch vie in world trade. Indeed, the value of goods and services which are exported from the Netherlands has amounted to nearly 50 percent of the Gross National Product and even exceeded that figure during 1974 and 1975.[118] Politically, it was the government's role to mediate between labor and management and, for most of the period between 1950 and 1965, there was general and broad-based support for the government's income policy. When conflict did arise, it was the government attempting to convince labor and management both that wage increases ran counter to the national interest.[119] Many of the circumstances which led to this wage policy also helped to bring about government intervention in the housing market in the Netherlands.[120]

Direct intervention by the Dutch government in the housing market was limited prior to World War II.[121] However, in 1940 rent controls were instituted as an emergency measure, a phenomenon first noted in many Western European countries as well as in the United States during the First World War.[122] In the Netherlands, rents were frozen at their 1940 level until 1951, a factor which precluded private capital from returning to the residential sector. Without incentives to respond to a severe housing shortage - estimated in 1947 to number 300,000 dwelling units - investors channeled their resources to other endeavors. Nevertheless, rent control could not be abandoned since national economic objectives required strict limits on wages and prices.

If the demand side of the ledger were modified by rent control, how was the crisis on the supply side dealt with? Given the emphasis on the reconstruction of the economy with particular attention paid to generating foreign exchange, the housing shortage could only be assuaged gradually. By 1949 the government determined that, if steady (albeit modest) efforts were made, the residential shortfall could be solved by 1965.[123] In the absence of an active private sector, the financial incentives utilized to generate residential construction were threefold. First, public funds were appropriated for the construction of rental housing by municipalities and by non-profit housing associations. Until 1969, this mode of financing accounted for about half of all housing units constructed in the Netherlands, divided more or less equally between housing associations and municipalities. A second source of government support to these sponsors is subsidies which enabled them to offer housing units at a monthly rental below break-even cost. And the third financial incentive to the housing market consists of construction subsidies to privately-financed dwelling units. Since 1969, these units have constituted the largest plurality in the Netherlands housing market and, along with rental dwellings constructed by building associations, comprise the type of housing characteristic of the post-1945 neighborhoods in this study (see Table 2.11). In addition, subsidies are provided for owner-occupied housing if acquisition cost and household income fall within certain limits. Privately-developed, non-subsidized housing, the 'free sector' of the housing market, represents unregulated access to housing units for the upper-income segments of the Dutch population. It has accounted for roughly 15 percent of post-1945 production.

In addition to stimulating housing investments, regional allocations of housing development are also a matter of government policy. Given the close relationship between regional development and national economic goals, addressing the housing shortage on a nationwide basis enabled proportional progress to be made, although by 1960 the numerical shortage remained identical to the 1947 figure of 300,000 units.[124] A very tight housing market continues to typify matters in the Netherlands as changing values, economic prosperity, and decreasing household size (a drop of 34 percent, from more than

TABLE 2.11 COMPLETED DWELLING UNITS BY DEVELOPER AND FINANCING CATEGORY IN THE NETHERLANDS: 1950-79

| | | Publically-Financed Units | | | | Privately-Financed Units | | | |
|---|---|---|---|---|---|---|---|---|---|
| | | Municipalities | | Building Soc. | | Subsidized | | Non-Subsidized | |
| Year | Units Completed | # | % | # | % | # | % | # | % |
| 1950a. | 47,300b. | | | | 45,600 | (96.4%) | | 1,700 | 3.6 |
| 1951 | 58,666 | | | | 54,600 | (93.1%) | | 4,100 | 6.9 |
| 1952 | 54,601 | | | | 52,000 | (95.2%) | | 2,600 | 4.8 |
| 1953 | 59,597 | | | | 57,900 | (97.2%) | | 1,700 | 2.8 |
| 1954 | 71,979 | 20,695 | 28.8 | 21,115 | 29.3 | 28,320 | 39.3 | 1,849 | 2.6 |
| 1955 | 64,019 | 16,932 | 26.4 | 14,911 | 23.3 | 29,626 | 46.3 | 2,550 | 4.0 |
| 1956 | 70,709 | 15,463 | 21.9 | 19,088 | 27.0 | 32,291 | 45.7 | 3,867 | 5.4 |
| 1957 | 91,738 | 23,051 | 26.1 | 24,147 | 26.3 | 39,500 | 43.1 | 5,040 | 5.5 |
| 1958 | 91,281 | 27,139 | 29.7 | 24,177 | 26.5 | 36,922 | 40.5 | 3,043 | 3.3 |
| 1959 | 85,505 | 24,436 | 28.6 | 22,912 | 26.8 | 35,922 | 42.0 | 2,235 | 2.6 |
| 1960 | 86,290 | 19,407 | 22.5 | 21,929 | 25.4 | 37,311 | 43.2 | 7,643 | 8.9 |
| 1961 | 84,533 | 17,385 | 20.6 | 16,712 | 19.8 | 30,568 | 36.1 | 19,868 | 23.5 |
| 1962 | 80,705 | 13,520 | 16.7 | 16,162 | 20.1 | 25,048 | 31.0 | 25,975 | 32.2 |
| 1963 | 84,307 | 15,164 | 18.0 | 19,204 | 22.8 | 21,909 | 26.0 | 28,030 | 33.2 |
| 1964 | 107,537 | 20,314 | 18.8 | 24,534 | 22.8 | 27,856 | 26.0 | 34,833 | 32.4 |
| 1965 | 119,919 | 24,999 | 20.8 | 29,856 | 24.9 | 28,411 | 23.7 | 36,653 | 30.6 |
| 1966 | 124,888 | 30,363 | 24.3 | 32,860 | 26.3 | 26,955 | 21.6 | 34,710 | 27.8 |
| 1967 | 130,402 | 33,278 | 25.5 | 39,131 | 30.1 | 26,540 | 20.3 | 31,453 | 24.1 |
| 1968 | 130,416 | 29,271 | 22.4 | 40,443 | 31.0 | 39,504 | 30.3 | 21,198 | 16.3 |

| | | | | | | | | | |
|---|---|---|---|---|---|---|---|---|---|
| 1969 | 134,643 | 23,386 | 17.4 | 39,065 | 29.0 | 53,620 | 39.8 | 18,572 | 13.8 |
| 1970 | 128,064 | 19,136 | 14.9 | 36,993 | 28.9 | 51,750 | 40.4 | 20,185 | 15.8 |
| 1971 | 152,919 | 12,214 | 8.0 | 54,135 | 35.4 | 64,320 | 42.1 | 22,250 | 14.5 |
| 1972 | 171,197 | 9,886 | 5.8 | 62,494 | 36.5 | 72,311 | 42.2 | 26,506 | 15.5 |
| 1973 | 172,794 | 9,042 | 5.2 | 63,505 | 36.8 | 68,572 | 39.7 | 31,675 | 18.3 |
| 1974 | 158,730 | 6,205 | 3.9 | 54,608 | 34.4 | 62,176 | 39.2 | 35,741 | 22.5 |
| 1975 | 126,515 | 5,188 | 4.1 | 40,683 | 32.2 | 54,467 | 43.1 | 26,177 | 20.6 |
| 1976 | 109,933 | 3,327 | 3.1 | 36,213 | 32.9 | 49,495 | 45.0 | 20,898 | 19.0 |
| 1977 | 113,629 | 2,856 | 2.5 | 35,682 | 31.4 | 47,844 | 42.1 | 27,248 | 24.0 |
| 1978 | 106,903 | 2,850 | 2.7 | 29,358 | 27.5 | 44,139 | 41.3 | 30,556 | 28.5 |
| 1979 | 88,094 | 2,578 | 2.9 | 23,822 | 27.0 | 32,788 | 37.2 | 28,906 | 32.9 |

a. No distinction was made between publically-financed housing and subsidized private housing during the years 1950-1953.

b. Totals may not agree with those of other sources due to some inconsistencies encountered in compiling data.

Sources: Compiled from Netherlands Central Bureau of Statistics, Statistical Yearbooks of the Netherlands (Staatsuitgeverij, The Hague, 1950-1981) and from J. Nycolaas, Volkshuisvesting; een bijdrage tot de geschiedenis van woningbouw en woningbouwbelied (Nijmegen: SUN, 1974), cited in Jan Van Weesep, Production and Allocation of Housing: the Case of the Netherlands. Geografische en Planologische Notities, No. 11 (Free University of Amsterdam, 1982), p.19

4.30 in 1947 to 2.86 in 1978) have relentlessly challenged government assumptions concerning the longevity of the housing shortage. As a result, the availability of housing is a major political issue whose most well-known manifestations are the occasional street battles between police and squatters in Amsterdam. On a less dramatic basis, this shortage situation has also prevailed in The Hague and environs where, for 20 percent of the sample in this study, the availability of a given housing unit constitutes the single most important determinant of the decision to occupy.

Despite record production levels by the mid-1960s, a decade which was to witness the share for housing attain a record high of almost 12 percent of all government expenditures, the housing shortage remained serious in the western Netherlands.[125] Although some optimists had forecasted the elimination of the shortfall by 1970, the government acknowledged in 1971 it was uncertain as to when it could proclaim that a sufficient supply of dwelling units existed. Production peaked in 1972 and although it did decline over the next few years, by 1975 it was officially pronounced that the housing crisis was history insofar as the absolute number of dwellings was concerned in that vacancies had reached a postwar high of 2.2 percent.[126]

The Netherlands is not the only nation confronting the conflict between housing production and trends regarding demographics and social expressions of growing affluence. In the United States 1984 data points to the burgeoning number of one-person households which, in that year, comprised half of the total, a figure which far exceeds that of nuclear families with one or two children, a category which accounts for only 35 percent of all households.[127] Questions of housing allocation in the United States are germane only insofar as the dynamics of supply and demand are concerned, but in the Netherlands, it has proven difficult to reconcile the absolute number of dwelling units with the changing needs of the population.

Mobility in the Netherlands is very much a function of local and national policy and, despite factional philosophical differences, it appears unlikely that sufficient incentives will be offered to non-moving tenants in order that dwelling units be matched with appropriate households.[128] Residents who have entered the housing market more recently or who have moved to a new or rehabilitated dwelling bear higher housing costs than those who have occupied their housing for a longer period of time. In 1977, more than 60 percent of renters with incomes in excess of 30,000 guilders spent less than 9 percent of their incomes on rent.[129] Despite sporadic increases in the 1950s, rents still lagged behind operating costs, the cost of living, personal incomes and far behind increases in construction costs and wages. This situation changed somewhat after 1960 as rents outpaced the overall cost of living but still failed to keep up with the increased costs of new construction.[130] Complicating this situation further is the high level of investment needed for infrastructure development required by the topography in the western Netherlands where most of the land is below sea level. For example, before building could proceed in a two square mile area annexed to West Amsterdam, 23 million cubic meters of sand were required for landfill.[131]

While filtering remains the implicit and unofficial American housing policy, its utility in the Netherlands remains an item for debate. On the one hand, there are those who believe that the free market should prevail in housing provided that the interplay of supply and demand is not constrained by shortage conditions. In addition, housing costs should reflect the quality of the dwelling unit. This general view characterizes the centrist and center-right administrations which held power from about 1960 until 1972. Unfortunately, the enduring shortage prevailed and government intervention in the market continued.[132]

In contrast, the philosophy of politicians on the left side of the spectrum calls for government intervention not because it is desirable that measures should be taken so that the free market can function, but because it is necessary to ensure that housing units and their costs be equitably distributed. To rely on the filtering process would result in outcomes reminiscent of American inner cities: environmental inequality, concentrations of low income groups, and ghettoization.[133]

One mechanism that has promoted housing redistribution is condominium conversion. Between 1947 and 1975 approximately 600,000 rental units entered the owner-occupied category by means of this process. This factor is largely responsible for the sizeable decrease in the number of privately-owned rental dwellings between 1956 (40 percent of the total) and 1975 (14 percent of the total).[134]

Ironically, the passage of the Horizontal Property Act in 1952 unwittingly facilitated this process by defining condominiums in much the same way as is done in the United States. Prior to that date, legal confusion prevailed in the general realm of collective ownership and the 1952 legislation was designed to clarify matters so that new condominium construction would be expedited. However, by not prohibiting conversions in existing buildings as a matter of law, if not policy, it became relatively simple to initiate the conversion process.[135]

The private sector responded enthusiastically to the condominium conversion option. This was particularly attractive to large institutional investors (insurance companies, pension funds, etc.) with holdings dating from the inter-war years. Although the condition of these properties was good, concern mounted over the growing disparity between rent increases and spiraling maintenance costs, especially after 1965. With profits declining, buildings already heavily depreciated, and other investments offering more attractive yields, it made sense to liquidate rental properties.[136]

A second impetus favoring conversions has arisen out of the vigorous growth in the number of one- and two-person households, a group which has often encountered problems locating suitable housing. This segment of demand, paired with the center-right government's encouragement of private initiative in the housing market,coincided perfectly with the investment strategies of those with large rental housing portfolios. These circumstances, therefore, account for the owner-occupied units within the Dutch sample described in this study.

The consequences of the condominium conversion phenomenon in the Netherlands are mixed. After 1965 major investors in the rental housing market found it advantageous, to lighten, if not liquidate their holdings of inter-war units. For those who purchased apartments at that time, costs were low and remain so for residents with continuous tenure. While this has severely diminished the rental stock and therefore limited public policy options, a positive stabilizing trend has been noted in pre-1940 residential areas.[137] In addition, the owner-occupants of these units have demographically and economically reinvigorated older neighborhoods and furthered prospects for their continued vitality. There are also tax benefits which accrue to owner-occupants in the form of tax deductions for mortgage interest and for major home ownership costs. In additon, some households qualify for insured mortgages with high loan-to-value ratios. And lastly, capital gains are not taxed although a small transfer fee is paid when a dwelling is sold. For many this latter point is of dubious value since housing prices have declined by 30 percent since the peak years of 1977-78.[138] As in the United States, rising interest rates, more cautious lending practices, and the world-wide recession have all quashed the equity appreciation aspirations of homeowners.

Nevertheless, Dutch housing policy after 1945 has been remarkably successful although not all problems have been solved. The number of dwelling units has outstripped increases in population although diminishing household size has offset some of this impressive numerical gain. The quality of accommodations has improved since the early 1960s. The floor area of publically-financed dwellings had increased by nearly 25 percent by 1973 and a comparable increment for subsidized private units has also been noted. The most dramatic increase in this dimension (38 percent) occurred in non-subsidized dwellings. In addition, plumbing, heating, and insulation standards have been signficantly upgraded.[139]

While an assessment of the workings of the entire Dutch housing market is beyond the scope of this discussion, there are a few issues that should be mentioned in conclusion since they continue to be of relevance, not only during the period in which this study's fieldwork was conducted (1978) but also for the decade of the 1980s. First, the quantitative housing shortage remains. It has been estimated that at least 110,000 units per year must be built between 1980 and 1990 if this is to be eliminated. As a result, housing choice will remain constrained and unavailability of dwelling units will continue to hamper mobility within the housing market. An indication of the magnitude of this problem is the fact that in 1981 over 350,000 households in the fourteen largest cities in the Netherlands are waiting for a suitable dwelling.[140] Second, with residential mobility constricted by inadequate supply, and with many more affluent households enjoying comparatively low housing costs, attempts to insure a more equitable distribution of dwellings will not succeed. Thirdly, while new production could be channeled in ways to correct existing imbalances, the prospects of a vigorous building industry in all western nations during the 1980s remain grim. Until the United States begins to confront the massive deficits which have launched an epidemic of high interest rates throughout the world, the construction industry will continue to suffer and the gap between housing needs and housing supply will continue to grow. In the Netherlands, the majority of the Dutch population has supported intertwined housing and regional development policies that have created prosperity and high standards of accommodation for much of the population. Certainly no nation has done more or studied these many-faceted problems so intently.

NOTES

[1] See Sam Bass Warner, Streetcar Suburbs, 2nd ed. (Harvard University Press, Cambridge, Mass., 1981); Peter J. Schmitt, Back to Nature; The Arcadian Myth in Urban America (Oxford University Press, New York, 1969); David Ward, Cities and Immigrants: A Geography of Change in Nineteenth Century America (Oxford University Press, New York, 1971); Joel Arthur Tarr, 'From City to Suburb: The "Moral" Influence of Transportation Technology', in Alexander Callow, (ed.), American Urban History (Oxford University Press, New York, 1973), pp.202-12.

[2] John C. Bollens and Henry J. Schmandt, The Metropolis, 4th ed. (Harper and Row, New York, 1982), p.5.

[3] Adna F. Weber, The Growth of Cities in the Nineteenth Century (Cornell University Press, Ithaca, New York, 1967), p.468.

[4] Moses Rischin, The Promised City: New York's Jews, 1870-1914 (Harper and Row, New York, 1970), p.79; Weber, The Growth of Cities in the Nineteenth Century, p.460.

[5] John F. Long, Population Deconcentration in the United States (U.S. Bureau of the Census, Washington, D.C., 1981), p.66.

[6] Ibid., p.68.

[7] James Heilbrun, Urban Economics and Public Policy, 2nd ed. (St. Martin's Press, New York, 1981), p.42.

[8] Long, Population Deconcentration, p.91.

[9] Basil G. Zimmer, 'The Urban Centrifugal Drift', in Amos H. Hawley and Vincent P. Rock, (eds.), Metropolitan America in Contemporary Perspective (Halsted Press/John Wiley, New York & London, 1975), p.52.

[10] Ibid.

[11] Heilbrun, Urban Economics, p.44; see also

Edgar M. Hoover and Raymond Vernon, Anatomy of a Metropolis (Harvard University Press, Cambridge, Mass., 1959), chapters 2 and 3.

[12] Heilbrun, Urban Economics, p.45.

[13] Zimmer, 'The Urban Centrifugal Drift', p.60.

[14] George Sternlieb and James W. Hughes (eds.), Shopping Centers: U.S.A. (Center for Urban Policy Research, New Brunswick, N.J., 1981), pp.2-3.

[15] Ibid.

[16] Bollens and Schmandt, The Metropolis, p.307.

[17] Ibid., p.245.

[18] Heilbrun, Urban Economics, p.282.

[19] John F. Kain and Joseph J. Perskey, 'Alternatives to the Gilded Ghetto', The Public Interest, no.12 (Winter 1969), p.76.

[20] Kenneth T. Jackson, 'Race, Ethnicity, and Real Estate Appraisal: The Home Owners Loan Corporation and the Federal Housing Administration', Journal of Urban History, vol.6, no.4 (August 1980), pp.419-52.

[21] Herbert J. Gans, The Levittowners (Vintage Books, New York, 1967), Chapter 14.

[22] Report of the National Advisory Commission on Civil Disorders (Bantam Books, New York, 1968).

[23] Barry Checkoway, 'Large Builders, Federal Housing Programmes, and Postwar Suburbanization', International Journal of Urban & Regional Research vol.4, no.1 (1980), pp.21-2.

[24] Cf. Robert C. Wood, 1400 Governments (Harvard University Press, Cambridge, Mass., 1961); Robert L. Bish, The Public Economy of Metropolitan Areas (Markham, Chicago, 1971); Charles Abrams, The Future of Housing (Harper and Brothers, New York, 1946); Martin Meyerson, et al., Housing People and Cities (McGraw Hill, New York, 1962); Sherman J. Maisel, Housebuilding in Transition (University of California Press, Berkeley, 1953).

[25] Jackson, 'Race, Ethnicity, and Real Estate', p.434.

[26] Checkoway, 'Large Builders', pp.32-3.

[27] Ibid., p.27.

[28] Ibid., p.28.

[29] Ibid., p.24.

[30] Larry Sawers and William K. Tabb (eds.), Sunbelt/Snowbelt: Urban Development and Regional Restructuring (Oxford University Press, New York & Oxford, 1984), p.8.

[31] AnnaLee Saxenian, 'The Urban Contradictions of Silicon Valley: Regional Growth and the Restructuring of the Semiconductor Industry', in ibid., p.175.

[32] Karl Belser, 'The Making of Slurban America', Cry California vol.5, no.4 (1970), p.1.

[33] Gene Bylinsky, 'California's Great Breeding Ground for Industry', Fortune, vol.89, no.6 (1974), pp.129-35, 216-24.

[34] Stanford Environmental Law Society, San Jose: Sprawling City (Environmental Law Society, Stanford, California, 1971), Chapter 2.

[35] Ibid.

[36] James M. Carney, 'How to Evaluate the Impacts of the Combined General Plans of the Cities of Santa Clara County, California', unpublished Master's Planning Report, San Jose State University, 1978, p.28.

[37] Santa Clara County Housing Task Force, Housing: A Call for Action (Santa Clara County Planning Department, San Jose, California, 1977), p.4.

[38] Santa Clara County Planning Department, Housing Characteristics, Cities, Santa Clara County, 1970 (Santa Clara County Planning Department, San Jose, California, 1971).

[39] Efraim Orni and Elisha Efrat, Geography of Israel (Israel University Press, Jerusalem, 1980), p.312-4.

[40] Ibid., pp.289-97.

[41] See Nadav Halevi and Ruth Klinov-Malul, The Economic Development of Israel (Frederick A. Praeger, New York, 1968), Chapter 2; and Daniel Shimshoni Israeli Democracy: the middle of the Journey (The Free Press, New York, 1982), Chapter 1.

[42] Orni and Efrat, Geography of Israel, pp.297-311.

[43] Haim Darin-Drabkin, Housing in Israel: Economic and Sociological Aspects (Gadish Books, Tel-Aviv, 1957), pp.19-35.

[44] Shikun, Housing in Israel (Workers Housing Co., Ltd., Tel-Aviv, 1965).

[45] Elisha Efrat, Urbanization in Israel (Croom Helm, London, 1984), pp.19-25.

[46] Ibid., pp.56-74.

[47] See Darin-Drabkin, Housing in Israel, p.28; and Great Britain, Department of Statistics, Statistical Abstract of Palestine: 1944-45 (Government Printer, London, 1946), p.21.

[48] Dan Soen, 'Israel's Population Dispersal Plans and Their Implications, 1949-74: Failure or Success?' Geo Journal, vol.1, no.5 (1977), pp.21-6.

[49] Darin-Drabkin, Housing in Israel, pp.36-41.

[50] It is estimated that during the 1948-49 Arab-Israeli conflict 550,000 Arabs migrated from

Israel to the West Bank and Gaza regions. See Fred M. Gottheil, 'On the Economic Development of the Arab Region in Israel' in Michael Curtin and Mordecai S. Chertoff (eds.), Israel: Social Structure and Change (Transaction Books, New Brunswick, N.J. 1973), pp.237-48.

[51] Darin-Drabkin, Housing in Israel, p.36.

[52] Amiram Gonen, 'The Suburban Mosaic in Israel', in D. H. K. Amiran and Y. Ben-Arieh (eds.), Geography in Israel (The Israel National Committee International Geographical Union, Jersualem, 1976), pp.163-86.

[53] Artur Glikson, 'Some Problems In Housing in Israel's New Towns and Suburbs', in Haim Darin-Drabkin (ed.), Public Housing in Israel: Surveys and Evaluations of Activities in Israel's First Decade (1948-1958) (Gadish Books, Tel-Aviv, 1959), pp.93-102; and Gonen, The Suburban Mosaic, pp.52-66.

[54] Israel Shaham, 'Public Housing In Israel', in J. S. Fuerst (ed.), Public Housing in Europe and America (John Wiley and Sons, New York, 1974), pp.52-66.

[55] Gonen, 'Suburban Mosaic', pp.172-4.

[56] See for example, Israel Ministry of Housing, Population and Building in Israel: 1948-73 (Jerusalem, 1975), pp.35-51; and Joseph Neipris, Social Welfare and Social Services in Israel: Policies, Programs, and Current Issues (The Hebrew University of Jerusalem, Paul Baerwald School of Social Work, Jerusalem, 1981), pp.98-102.

[57] Iris Graizer, 'Spatial Patterns and Residential Densities in Israeli "Moshavot" in Process of Urbanization', Geo Journal, vol.2, no.6 (1978), pp.533-7.

[58] Ibid.

[59] Alexander Berler, et al., Urban-Rural Relations in Israel: Social and Economic Aspects (Settlement Study Center, Rehovot, 1970).

[60] See Ibid. and Gonen, 'Suburban Mosaic', pp.181-2.

[61] Efrat, Urbanization in Israel, pp.8-10.

[62] Michael Romann, 'Jews and Arabs in Jerusalem', The Jerusalem Quarterly, no. 19 (Spring 1981), pp.23-46.

[63] Ibid.

[64] Gonen, 'Suburban Mosaic'.

[65] Efrat, Urbanization in Israel, pp.67-9; and Shimshoni, Israel Democracy, pp.309-14.

[66] Efrat, Ibid., p.9.

[67] Myron J. Aronoff, Frontiertown: The Politics

*of Community Building in Israel* (Manchester University Press, Manchester, 1974), pp.19-25.

[68] Arie Shachar, 'New Towns in a National Settlement Policy', *Town and Country Planning*, vol.44, no.2 (1976), p.83.

[69] *Ibid.*

[70] *Ibid*; Shimshoni, *Israeli Democracy*, pp.313-4.

[71] E.A.J. Johnson, *The Organization of Space in Developing Countries* (Harvard University Press, Cambridge, Mass., 1970), pp.296-310.

[72] *Ibid.*

[73] Efrat, *Urbanization in Israel*, p.141.

[74] Johnson, *The Organization of Space*, pp.301-5.

[75] Shachar, 'New Towns', pp.83-7.

[76] Shimshoni, *Israeli Democracy*, pp.117-9.

[77] Efrat, *Urbanization in Israel*, p.150.

[78] *Ibid.*

[79] Johnson, *The Organization of Space*, pp.302-8.

[80] Artur Glikson, 'Urban Design in New Towns and Neighborhoods', *Ministry of Housing Quarterly* (December 1967), pp.45-51.

[81] Israel, Central Bureau of Statistics, *Statistical Abstract of Israel: 1978* (Sivan Press, Ltd., Jerusalem, 1978) p.48.

[82] Shachar, 'New Towns', p.84.

[83] See, for example, Myron J. Arnoff, *Frontiertown: The Politics of Community Building in Israel*, Chapter 1; Alexander Berler, *New Towns in Israel* (Israel Universities Press, Jerusalem, 1970), pp.66-79; Yochanan Comay and Alan Kirchenbaum, 'The Israeli New Town: An Experiment of Population Redistribution', *Economic Development and Cultural Change*, vol.22, no.1 (1974), pp.124-34; and Artur Glikson, 'Some Problems of Housing in Israel's New Towns and Suburbs'.

[84] Naomi Carmon and Bilha Mannheim, 'Housing Policy as a Tool of Social Policy', *Social Forces*, vol.58, no.2 (1979), pp.336-51; and Comay and Kirchenbaum, *Ibid*.

[85] Efrat, *Urbanization in Israel*, pp.69-70.

[86] Soen, 'Israel's Population Dispersal Plans', p.21; and Shimshoni, *Israeli Democracy*, pp.378-81.

[87] Johnson, *The Organization of Space*, pp.309-10.

[88] Efrat, *Urbanization in Israel*, Chapter 5.

[89] Although the public sector was traditionally the major supplier of housing in Israel, the private sector has dominated the housing market since the mid 1970s due to changes in government policy and consumer demand. See Eli Borukhof, Yona Ginsberg and Elia Werczberger, 'Housing Prices in Housing

Preferences in Israel' Urban Studies, vol.15, no. 2 (1978), pp.187-200; and Stuart Gabriel and Ilan Maoz, 'Cyclical Fluctuations in the Israel Housing Market', Center for Real Estate and Urban Economics Graduate School of Business, University of California, Berkeley, 1983).

[90] Efrat, Urbanization in Israel, pp.70-1.

[91] P. W. Klein, 'The Foundations of Dutch Prosperity', in Richard T. Griffiths, (ed.), The Economy and Politics of the Netherlands Since 1945 (Martinus Nijhoff, The Hague, 1980), p.3.

[92] Ibid., p.4.

[93] Ibid., pp.6-7.

[94] Ibid., p.11.

[95] Netherlands Minister of Economic Affairs, Memorandum on the Industrialization of the Netherlands (September 1949), cited in James G. Abert, Economic Policy and Planning in the Netherlands, 1950-1965, (Yale University Press, New Haven & London, 1969), p.6.

[96] For more on the Dutch political system, see A. Lijphart, The Politics of Accommodation: Pluralism and Democracy (University of California Press, Berkeley & Los Angeles, 1968).

[97] Abert, Economic Policy, p.6; Klein, 'Foundations of Dutch Prosperity', p.8.

[98] W. Brand, 'The Legacy of Empire', in Griffiths, p.261.

[99] Ibid., p.265.

[100] The Netherlands Ministry of Housing and Physical Planning, The Relationship Between Physical Planning Policy and Economic Policy, (Staatsuitgeverij, The Hague, 1977), p.1.

[101] David Pinder, The Netherlands, (Westview Press, Boulder, Colo., 1976), p.63.

[102] Ibid., p.64.

[103] G. A. van der Knaap, 'Sectoral and Regional Imbalances in the Dutch Economy', in Griffiths, p.117.

[104] Ibid., p.121.

[105] Ibid., p.128. See also Peter Hall, The World Cities, 2nd ed. (McGraw Hill, New York, 1977), Chapter 4; Pinder, Chapters 5-6; G. R. P. Lawrence, Randstad Holland, (Oxford University Press, London, 1973); and Gerald L. Burke, Greenheart Metropolis: Planning the Western Netherlands (St. Martin's Press, New York, 1966).

[106] See D. J. Van de Kaa, 'Population Prospects and Population Policy in the Netherlands, Netherlands Journal of Sociology, vol.17 (1981), pp.73-91.

[107] Hall, The World Cities, p.104.

[108] The Netherlands Ministry of Housing and Physical Planning, Summary of the Report on Urbanization in the Netherlands (Staatsuitgeverij, The Hague, 1976), p.10.

[109] Ibid., p.5.

[110] Ibid., p.12.

[111] Steven Hamnett, 'The Netherlands; Planning and the Politics of Accommodation', in David H. McKay, (ed.), Planning and Politics in Western Europe (St. Martin's Press, New York, 1982), p.135; The Netherlands Ministry of Housing and Physical Planning, Summary of the Orientation Report on Physical Planning, (Staatsuitgeverij, The Hague, 1974), p.4.

[112] Brian J. L. Berry, Urbanization and Counterurbanization, (Sage, Beverly Hills & London, 1976).

[113] David Vining and Thomas Kontuly, 'Population Dispersal from Major Metropolitan Regions: An International Comparison', International Regional Science Review, vol.13, no.1 (1978), p.49. See also A. J. Fielding, 'Counterurbanization in Western Europe', Progress in Planning, vol.17, Part 1 (1982).

[114] Hamnett, 'The Netherlands: Planning and the Politics of Accommodation', p.138.

[115] McKay, Planning and Politics in Western Europe, p.189.

[116] David H. McKay, 'Planning in the Mixed Economy: Problems and Prospects', in McKay, Planning and Politics in Western Europe, p.178.

[117] Ibid.

[118] P. de Wolff and W. Driehuis, 'A Description of Post War Economic Developments and Economic Policy in the Netherlands', in Griffiths, The Economy and Politics of the Netherlands Since 1945, p.18.

[119] Abert, Economic Policy, p.71.

[120] See Abert, Chapter 2, and de Wolff & Driehuis, op. cit.

[121] See Jan Van Weesep, Production and Allocation of Housing: the Case of the Netherlands, Geografische en Planologische Notities, No.11, (Free University of Amsterdam, Geografische en Planologisch Instituut, Amsterdam, 1982), pp.1-16; Donald I. Grinberg, Housing in the Netherlands 1900-1940, (Delft University Press, Delft, 1977).

[122] Louis M. Rea and Dipak K. Gupta, An Economic and Legal Analysis of Rent Control (Institute of Public and Urban Affairs, San Diego State University, San Diego, 1982); Hugo Priemus, 'Rent

Control and Housing Tenure', Planning and Administration, vol.9, no.2 (1982), pp.29-46.

[123] Van Weesep, Production and Allocation of Housing: the Case of the Netherlands, p.17. See also the same author's 'Intervention in the Netherlands: Urban Housing Policy and Market Response', in Urban Affairs Quarterly, vol.19, no.3 (1984), pp.329-53.

[124] Van Weesep, Production and Allocation, pp.18, 21.

[125] Ibid., pp.27-8.

[126] Ibid., and Van Weesep, 'Urban Housing Policy'.

[127] John B. Allen, 'The Next Seven Years: A Real Estate Perspective', Grubb & Ellis Investor Outlook, vol.4, no.2 (1984), p.3.

[128] For more on this question see W. A. V. Clark & P. C. J. Everaers, 'Public Policy and Residential Mobility in Dutch Cities', Tijdschrift voor Economische en Sociale Geografie, vol. 72, no. 6 (1981), pp.322-33.

[129] Van Weesep, Production and Allocation, p.50.

[130] Ibid., p.49.

[131] Haim Darin-Drabkin, Land Policy and Urban Growth, (Pergamon Press, Oxford & New York, 1977), p.354.

[132] See Van Weesep, Production and Allocation, Chapter 4.

[133] H. Kombrink, 'Huidig doorstromingsbelied is te weining op behoeften afgestemd', Bouw, no.25 (1978), pp.42-44, cited in Van Weesep, p.32.

[134] F. van de Bergh, 'De koers is on maar het schip zel stranden', Bouw, no.24 (1978), 17-21, cited in Van Weesep, p.43.

[135] M. W. A. Maas, 'Condominium Conversion in Pre-War Neighborhoods: An Urban Transformation Process in Dutch Cities', Tijdschrift voor Economische en Sociale Geografie, vol.75 (1984), p.37.

[136] J. Van der Schaar, Sektorindeling en woningmarkt processen (Staatsuitgeverij, The Hague, 1979), cited in Maas, p.38.

[137] Ibid., p.45.

[138] Van Weesep, Production and Allocation, pp.22, 47.

[139] Ibid., p.43.

[140] Ibid., p.54.

# CHAPTER 3

# A STUDY OF THE QUALITY OF LIFE OF THE SUBURBAN ENVIRONMENTS IN THREE COUNTRIES

The previous chapters have established that metropolitan decentralization and suburban growth are hallmarks of advanced industrialized societies. In addition these processes and related policies have been examined in the national contexts of three industrialized countries: the United States, Israel, and the Netherlands. In order to obtain a more detailed cross-national study, this chapter presents an examination of levels of satisfaction with suburban environments in metropolitan areas representative of these industrialized countries: the San Jose region; the Tel-Aviv and Jerusalem areas; and The Hague and environs.

## A. METHOD

This study tries to measure levels of satisfaction with various dimensions of suburban life for women and their families residing in a range of social and physical environments representative of metropolitan life in the United States, Israel and the Netherlands. The dimensions of suburban life examined are variables depicting satisfaction with some of the major aspects of the quality of metropolitan life, such as housing and community services. Also relevant in this study are demographic and environmental characteristics (independent variables) such as education, income, housing density, and design features, which seem to be related to such levels of satisfaction (dependent variables). That is, expected relationships among variables are tested to help explain the variations anticipated in levels of satisfaction of suburban life. In addition, exploratory studies are conducted which pertain to aspects of the suburban environment such as locational preferences of suburban residents.

In particular, the influence of four theory-related clusters of independent variables are examined (social class, subcultural, life cycle, and environmental) with respect to four sets of dependent quality of life variables (housing environment, community services, social patterns, psychological well-being). The research design for this study is presented in Figure 3.1.

Questions about each variable or measures of satisfaction were devised and put into the form of a questionnaire schedule. As Appendix A indicates, these questions were designed to yield an ordinal satisfaction score in accordance with widely-used, Likert-type scales such as those employed by Lansing, et al.[1] in their study of satisfaction of planned residential environments, Anderson, et al.[2] in their examination of satisfaction of low- and moderate-cost housing, Zehner[3] in his research on the quality of life in new communities, and Fischer[4] in his study of personal networks in town and city. Thus, a score can be obtained for the 'personal and family privacy' variable by asking:

Is it hard to find a place to be by yourself in the house?

| 1 | 2 | 3 | 4 | 5 | 6 | 7 |
|---|---|---|---|---|---|---|
| Always | | | Not sure | | | Never |

In Phase I of this study (Summer-Fall 1976) the questionnaire was constructed, pre-tested and revised. During Phase II, (1977) the questionnaire was administered to approximately 800 women residing in eight socially and economically-matched, middle-income suburban neighborhoods in the San Jose metropolitan area. In Phase III (1978) translated versions of this questionnaire were employed in a similar manner in the metropolitan areas of Tel-Aviv, Jerusalem and The Hague.

1. Dependent Variables

Characteristics of suburbia chosen for this study were intended to yield a comprehensive set of indexes concerning the satisfaction of suburban life - physical, social and psychological. While it appears that no broadly accepted theory about the quality of metropolitan life exists,[5] the dimensions of suburban life were chosen for this study from conceptualizations about the quality of urban life most frequently discussed by urban scholars.[6] As shown in Figure 3.1, these dimensions of suburban

FIGURE 3.1 RESEARCH DESIGN

Independent Variables (IV):

a. Social Class Influences
    1. Income
    2. Education
    3. Occupational Status

b. Subcultural Influences
    1. Ethnic Identity
    2. Regional Origin
    3. Length of Residency

c. Life Cycle Influences
    1. Age
    2. Marital Status
    3. Number and Age of Children

d. Suburban Environmental Influences
    1. Residential Density
    2. Household Distance from the Central City Center
    3. Age of Neighborhood
    4. Distance to Work
    5. Population Size of Political Unit
    6. Design and Site Plan Characteristics

Dependent Variables (DV):

a. Housing Environment Satisfaction
    1. Personal and Family Privacy
    2. Size and Arrangement of the Housing Unit and Lot
    3. Economic and Functional Responsibilities of Household
    4. House and Neighborhood Appearance

b. Community Service Satisfaction
    1. Parks/open space
    2. Schools
    3. Police/security
    4. Child Care
    5. Transportation
    6. Entertainment and Culture

c. Social Patterns Satisfaction
    1. Friendships
    2. Group participation
    3. Sense of belonging

FIGURE 3.1 RESEARCH DESIGN (continued)

d. Psychological Well-Being
   1. Fullness vs. Emptiness of Life
   2. Social Respect
   3. Personal Freedom
   4. Companionship vs. Isolation
   5. Tranquility vs. Anxiety
   6. Self-Approving vs. Guilty
   7. Self-Confidence
   8. Elation vs. Depression

Where $(DV)_i = a + b_j(IV)_j + b_k(IV)_k +$

and $(DV)_i$ = the $_i$th Dependent Variable.

$(IV)_j$, $(IV)_k$ = the $_j$th and $_k$th Independent Variables

a and b are coefficients

---

life were grouped into four clusters of dependent variables concerned with overall satisfaction of housing environment, community services, social patterns, and psychological well-being. For example, the variables chosen as specific indexes of satisfaction with the housing environment are personal and family privacy; size and arrangement of the housing unit and lot; economic and functional responsibilities of household; and house and neighborhood appearance. In turn, each of these specific indexes of satisfaction is comprised of detailed measures of satisfaction, each related to a question in the questionnaire. As Figure 3.2 indicates, the detailed measures associated with the specific index of 'personal and family privacy' are concerned with unwanted visits; audio and visual privacy; and indoor and outdoor privacy. Since the relative importance of these measures is uncertain, the scores of each of these measures are weighed equally in constructing the specific index of satisfaction of personal and family privacy, and in turn, the specific indexes are weighed equally in building the overall index of satisfaction of the housing environment.

The other indexes of suburban life satisfaction were constructed in a similar manner, with one exception: the specific indexes of psychological well-being were largely based on a particular scale which attempts to measure aspects of mental well-being considered to be most appropriate for the

FIGURE 3.2 CONSTRUCTION OF OVERALL INDEX OF SATISFACTION

| | | | | | |
|---|---|---|---|---|---|
| Overall Index of Satisfaction | Housing Environment | | | | |
| Specific Indices of Satisfaction | (1) Personal and Family Privacy | (2) Size and Arrangement of Housing Unit & Lot | (3) Economic and Functional Responsibilities | (4) House and Neighborhood Appearance | |
| Detailed Measures of Satisfaction | (a) Unwanted Visits | (b) Audio Privacy | (c) Visual Privacy | (d) Indoor Privacy | (e) Outdoor Privacy |

where:

$$(1) = \frac{(a) + (b) + (c) + (d) + (e)}{5}$$

and:

$$\text{Housing Environment Score} = \frac{(1) + (2) + (3) + (4)}{4}$$

study of a non-pathological population - the Personal Feeling Scale by Wessman and Ricks.[7] This scale is similar in content to the semantic differential scales used in constructing 'Index of General Affect' by Campbell, Converse and Rodgers in their attempt to develop a qualitatively detailed way to measure overall life satisfaction or happiness.[8] Other measures of dimensions of the quality of suburban life, such as those for the housing environment, attempt to deal with an individual's satisfaction with a certain aspect of life based on the conceptual model shown in Figure 3.3, which relates satisfaction with an attribute being assessed to the perception of that attribute, and the standard against which the attribute is judged. In contrast, the scale for psychological well-being was designed primarily as a hedonic scale to measure the relative 'happiness-unhappiness' of an individual. While it has been argued that happiness measures are less stable than those for satisfaction due to short-term emotional fluctuations on which happiness measures are based,[9] the psychological scale was shown to be fairly stable over time.[10] We believe that the emotional-oriented 'happiness' measures reflect more deeply an individual's inner sense of well-being than do cognitive judgmental-oriented 'satisfaction' measures.[11]

Apparent differences between measures of happiness and satisfaction may not make an overwhelming difference as it was shown that global measures of happiness and satisfaction have a substantial overlap ($r = 0.50$) and that semantic differential scales similar in content to the ones employed could have very similar correlations with measures of both happiness and satisfaction.[12] Thus, scales designed to estimate either levels of happiness or satisfaction measure similar attributes and it is likely that the scale for psychological well-being will reflect both an individual's happiness and feelings about overall life satisfaction.

The construction of all overall indexes of satisfaction is presented in Appendix B.

## 2. Independent Variables

Based on theories and previous research concerning the influences of social class, subculture, life cycle, and the environment, independent variables representing crucial demographic and environmental characteristics were chosen as the most likely

FIGURE 3.3 CONCEPTUAL MODEL OF SATISFACTION

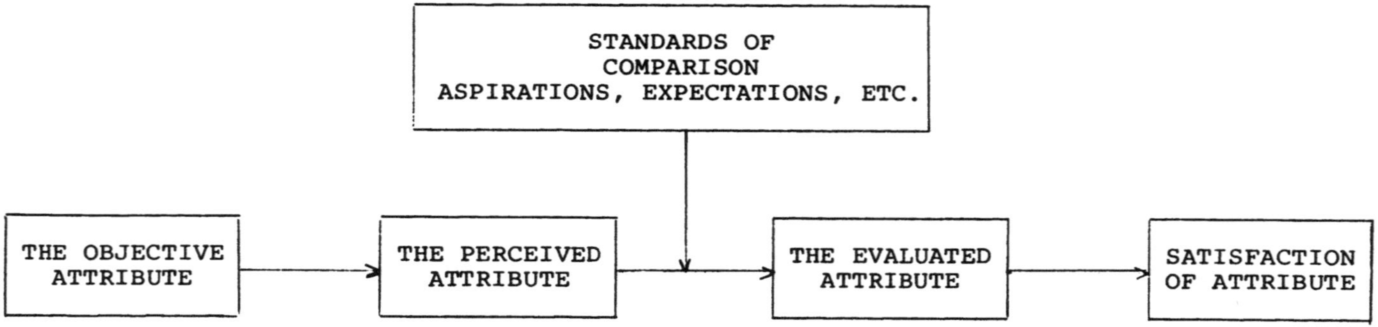

Source: Angus Campbell, Philip E. Converse and Willard L. Rodgers, _The Quality of American Life_ (Russell Sage Foundation, New York, 1976), p.13

indicators of changes in the satisfaction levels of dependent variables. These independent variables, which are presented in Figure 3.1, are grouped into four theory-related categories: social class influences; subcultural influences; life cycle influences; and suburban environmental influences. A discussion is presented of the independent variables associated with each of these categories as well as of their relationships with dependent variables.

Because the literature that has been reviewed is often contradictory in its predictions and has been based on research conducted over a wide range of time and circumstances, and because of the exploratory nature of this study, it is not appropriate to posit formal hypotheses. Rather, only general and tentative expectations of the relationships among the variables examined can be stated.

a. Social Class Influences

It has been argued that the independent variables of income, education and occupational status are related to social class influences which affect the satisfaction of dependent variables, regardless of the residential environment.[13] Accordingly, a testing of these variables will be used to examine the validity of social class theory.

1. Income

Although it is clear that increases in family income purchase increments of housing quality, it has also been shown that such increases allow families to spend a smaller percentage of their income on housing.[14] Accordingly, higher income families can use a greater percentage of their incomes to purchase other goods and services (e.g., collective goods, transportation, child care, medical care, education) than lower income households. In addition, there is evidence suggesting that increases in family income are positively associated with improved overall life satisfaction and mental health.[15] We, therefore, expect a positive relationship between family income and all dependent satisfaction variables.

2. Education

Researchers have shown that households with high levels of education (years) are willing to spend more of their income on housing[16] and on a planned residential environment rich in public facilities than do less-educated households with the same income.[17] Thus, highly-educated households are more likely to be found in housing environments with access to superior public facilities, particularly schools.

However, since such households may have more sophisticated functional, social, and aesthetic tastes for housing, community facilities and friendship patterns than those generally provided by the market in most relatively new suburban areas, there could be considerable dissatisfaction with many aspects of their environments.[18] If true, this in turn could lead to socialization problems and mental stress. Yet, other studies found a strong positive relationship between level of education and social satisfaction.[19] Further, an American national survey found that, in general, the most educated women have the greater options in life, such as professional development; and, for employed married women, there is a positive relationship between level of education and overall life satisfaction.[20] Given such findings, it is anticipated that increases in the educational level of women will be associated positively with all dependent satisfaction variables.

3. Occupational Status

Since occupational status is often highly correlated with years of education, it could be excluded as a separate indicator of social class.[21] However, occupational status was included in this study because of its potential relation to psychological well-being. Since a woman is more likely to experience a greater discrepancy between occupational status and level of education than does a man,[22] it seems likely that an increase in a woman's occupational status would have a more positive impact on mental health than an increase in education.[23] Indeed, studies have shown that employment status can have an important impact on overall life satisfaction for married women.[24] Therefore, a positive relationship between occupational status and psychological well-being variables is expected.

b. Subcultural Influences

Related to social class theory are subcultural influences concerned with ethnic or regional ties.[25] The subcultural theory suggests that these ties independently influence preferences for and satisfaction with metropolitan environments.[26] In order to test this general expectation, the independent variables of ethnic identity, regional origin, and length of residency are examined.

1. Ethnic Identity

Considerable evidence indicates that households with strong ethnic ties, particularly if combined with working class ties, often have a difficult time adjusting to the low density, low interaction level of a typical middle class American suburb.[27] It is supposed that a portion of the metropolitan population in the United States, Israel, and the Netherlands is comprised of 'ex-ethnics' - individuals who are accepted and function in middle class metropolitan life but who grew up in more culturally distinct environments (e.g. Italians and Irish in the United States, Oriental Jews in Israel, immigrant families in the Netherlands) and who miss the intensity and character of their culture of origin.[28] If this is true, 'ex-ethnics' may find many physical and social aspects of their suburban environments uncomfortable, dull and generally unsatisfying. Such patterns of negative relationships are expected to carry over into the realm of mental health and that ethnic identification will be negatively related to all dependent variables.

2. Regional Origin

Since the end of World War II, the San Jose Standard Metropolitan Statistical Area (SMSA) has been one of the most rapidly growing metropolitan regions in the United States. Most of this growth has been due to immigration from other regions. Hence, much of suburbia in San Jose, and the expanding American sunbelt in general, is inhabited by adults who spent their formative years in other regions. Although less pronounced, partly due to a smaller geographic scale, other industrialized nations, such as the Netherlands, have experienced similar inter-regional mobility; and still others, such as Israel, have accommodated dramatic numbers of immigrants from many distant countries.[29]

When the household head is from a large urban area where there is a high population density, that individual and family may develop greater social distance from the public; as a result, social activities are more accessible to sophisticated households.[30] However, in rural areas, the number of people one meets is limited but the inclination to know them may be greater; society-at-large is probably not a significant factor and the family will tend to be less sophisticated in dealing with social complexities and subcultural variety.[31] Therefore, it may be easier for individuals from large metropolitan areas to adjust to a new housing environment in an urban region than those from smaller communities.

It has also been argued that women moving beyond commuting distance from their origins[32] to areas with unfamiliar physical and social interaction patterns have substantial difficulties adjusting to a new environment.[33] Therefore, it is expected that all satisfaction dependent variables would be positively related to the degree of urbanness (e.g., population size) of the community in which a respondent's formative years were spent and negatively related to the distance of that community to the metropolitan area of current residence.

3. Length of Residency

Although it has been claimed that some forms of loneliness (familial, existential, emotional) may not decrease easily with length of residence,[34] other studies have shown that mutual assistance among neighbors increases with the length of time a family has lived in one place.[35] This latter process may be especially true for residents having strong ethnic or subcultural ties.[36] It also seems likely that increased familiarity with the housing environment, community services, and friendship patterns generates more overall satisfaction with the passage of time.[37] Hence, it is anticipated that length of residency will be positively related to all dependent variables.

c. Life Cycle Influences

The life cycle theory suggests that one's preferences for and satisfactions with the housing environment are related to the stage of one's life cycle, rather than to the environment itself.[38] For example, households in the child-rearing stage are expected to be more satisfied with outlying, relatively low-density housing than with inlying, high-density residential environments.[39] The independent variables associated with life cycle, age, marital status, and number and age of children, will be used to test this general conceptualization.

1. Age

Researchers have found that the younger the head of household, the more likely that home purchase has resulted in economic problems,[40] which in turn have been identified as the major cause of stress in newly-established suburbs.[41] While some research indicates that women at the end of the child-rearing life cycle stage (35-50) are more likely to participate in the labor force,[42] and enjoy more economic and psychic benefits than do younger women with children, a considerable amount of evidence indicates that mental health may decline in response to increases in age for the general population.[43] Therefore it is expected that age will vary positively with satisfaction with the housing environment and negatively with satisfaction variables concerned with psychological well-being.

2. Marital Status

Although it has been shown that many unmarried women achieve high levels of economic independence,[44] there is considerable evidence that single women who are heads of households are likely to have substantially more problems than their married counterparts concerning income,[45] household responsibilities,[46] mental health and social patterns,[47] and the enjoyment of community services.[48] Because of the extent of problems associated with unattached women in what may be a couple-oriented suburban world, it is expected that women who are single heads of households will generally experience lower scores for all dependent satisfaction variables than will married women.

3. Number and Ace of Children

Other studies have shown that the presence of children is directly related to neighboring and to participation in formal organizations.[49] In addition, there is evidence that the greater the number of children, the greater the degree of desire for low density housing;[50] and the older the children, the greater the degree of dissatisfaction with a low density environment.[51] At the same time, the increasing desire of women to participate in the labor force and other non-household related activities[52] would suggest the opposite set of relationships: an increase in satisfaction with a decrease in the number of children and an increase in age of children.[53] Therefore, it is expected that all dependent satisfaction variables will vary negatively with the number of children and positively with the mean age of all children.

d. Suburban Environmental Influences

Environmental theories suggest that the physical environment has considerable influence on an individual's level of satisfaction with that environment.[54] Despite evidence that might challenge this concept, the significance of these environmental influences will be examined by testing variables associated with characteristics of the residential environment: residential density; household distance from the central city center; age of neighborhood; distance to work; population size of political unit; and design and site plan characteristics.[55]

1. Density

Studies have shown that housing satisfaction can vary inversely with residential density.[56] While conflicting findings exist about density,[57] it appears that the lower the density, the more lifestyles will tend to emphasize the role of the nuclear family[58] and the greater the degree of social participation.[59] Further, the lower the density the more likely one will find adaptive behavior based on specialized interests.[60] As density increases it becomes more probable that a female head of household will (a) be employed, (b) have fewer children, (c) be divorced or separated, (d) be more mobile, (e) engage in less neighboring, and (f) engage in less community or civic activity.[61] In addition, as density increases, the more likely one will find restrictions on recreational and leisure activity.[62] Since low density housing is often a function of income and status,[63] it is believed that housing satisfaction will increase as density increases, and that lower densities are associated with greater friendship and organizational participation for married women. Employed single heads of households are likely to find these opportunities less advantageous and therefore a low density situation will probably be less satisfying for them. It is anticipated that density will be inversely related to the dependent satisfaction variables concerned with the housing environment and social patterns for married women, and the reverse relationship will prevail for women who are single heads of households.

2. Household Distance from the Central City Center

Clearly, the central cities of metropolitan areas in the United States and other industrialized nations are declining as regional employment foci.[64] Yet, cities still remain dominant centers for cultural, educational, and governmental activites as well as for health and other important services. While it is true that these non-business activities have also decentralized, they tend to be relatively less-dispersed throughout the lower density suburban areas.[65] Since it has been shown that women have less access to the automobile than do men,[66] and the quality of public transportation in suburban areas is generally limited,[67] it seems likely that women living in outlying suburban areas would have less access to community services of the central city as well as to their social and psychological benefits than do women who live in inlying neighborhoods.[68] It is therefore expected that household distance from the central city center (miles) will vary negatively with the satisfaction variables concerned with community services, social patterns, and psychological well-being.

3. Age of Neighborhood

Residents of established areas may have a clearer sense of social reality than those in newer areas, and women are most likely to be sensitive to this situation's impact.[69] It seems probable that older neighborhoods would find more cohesive and stable social organization in terms of the quantity and quality of friendships and in the choice of organizational affiliations. To the extent that older neighborhoods are likely to have residents with a greater average length of residency than those in newer developments, they should generate much of the physical, social and psychological benefits attributed to individual length of residency.[70] It is therefore anticipated that age of neighborhood will have a positive association with all dependent satisfaction variables.

4. Distance to Work

Researchers have documented that the journey to work is among the most important trips (as a percentage of all trips) emanating from households in metropolitan areas of industrialized nations.[71] Indeed, employment location in the United States has been found to be the major force determining residential location for non-minority groups.[72] That is, once the employment location of the primary earner is established, a housing unit is found within the commuter-shed of that location. It has been argued that most families in the lifecycle phase of child-rearing prefer to consume more inexpensive land in outlying suburban areas in exchange for less accessibility - higher cost and time loss for commutation to work - than their inlying counterparts.[73] This appears to be the case even for individuals who both live and work in the suburbs.[74]

The question we would like to address is whether increases in commutation time affect levels of satisfaction with aspects of the suburban environment. For example, it is known that the journey to work can often lengthen the work day by as much as 20 percent and therefore cut deeply into time otherwise spent on social and leisure activities,[75] or conflict with home roles.[76] It has also been shown that longer journeys to work not only lessen overall satisfaction for a commuter but also for his or her spouse, due to excessive time separation.[77] It is therefore anticipated that the travel time to work of a woman and/or her spouse will vary negatively with satisfaction with community services, social patterns and with psychological well-being.

5. Population Size of Political Unit

While factors other than political have dictated massive shifts of population from central cities to the suburbs,[78] the net political effect of suburbanization has been the fragmentation and decentralization of governmental units.[79] This phenomenon favors those whose personal resources are sufficient to solve their own immediate problems[80] because the smaller the unit, the smaller a range of public goods that unit can furnish.[81] Thus, the suburbs provide the setting where particularistic goals and values may be vigorously pursued.[82] Expressed in economic terms, the higher the socio-economic status of a unit of local government, the smaller its population is likely to be.[83] Further, it has been argued that political fragmentation is more efficient since each (relatively affluent) individual may choose and maximize the package of public services deemed most desirable.[84] Therefore, it is expected that the population size of a political unit will vary inversely with satisfaction on all community services variables.

6. Design and Site Plan Characteristics

Mass-produced suburban housing in the

industrialized nations has come under fire from a number of sources for environmental and social reasons.[85] Yet, as more people acquire a suburban background, two related trends are likely to emerge. First, those who have lived in the suburbs are more likely to want to continue to live there.[86] Second, these 'suburbanized' individuals will probably tend to be more localized in their contacts.[87] It has been shown that this community-centered population is attracted to planned communities.[88] These individuals are generally upper-middle class and seek protection from unplanned change, physical as well as social, as well as access to shared facilities.[89]

For example, residents of Westlake Village, a planned community near Los Angeles of mixed housing types and densities, conform to the characteristics cited in the preceding discussion. Seventy-five percent lived previously in suburbs of Los Angeles or other large cities; only 4 percent came from central cities. The population is solidly upper-middle class in income and occupation. Previous amenities characteristic of suburban life (schools, social homogeneity) are taken for granted. Indeed, the chief concern of the residents of Westlake Village is a master plan which guarantees open space and environmental amenities which include recreational facilities.[90] While different in housing type and density from their American counterparts, highly planned suburban developments in other industrialized nations have also been very successful.[91] Therefore, it is expected that the greater the degree of planned characteristics in a housing development, the greater the level of satisfaction on all dependent variables.

## B. SUBJECTS AND SAMPLE

### 1. The United States

Utilizing census tract and block statistics, and preliminary field surveys, eight middle and upper-middle class suburban neighborhoods (similar in terms of social and economic characteristics) were identified which typify the range of suburban development in the San Jose metropolitan area (Santa Clara County) and which ensure a substantial variation of most independent variables concerned with the environment. The social, economic, and physical characteristics employed to identify study neighborhoods were independent variables (discussed above) such as years of school completed (for the population over 25 years old), annual household income ($10,000 to $40,000 in 1975), household distance from the central city center (miles), residential density (households per acre), and design and site plan characteristics (planned environments vs. undifferentiated suburban development).[92] Where feasible, data from the Special Census of the San Jose SMSA conducted in 1975 were incorporated in this selection process.[93]

Subjects for this study were randomly selected women from each neighborhood. As we are primarily interested in women in the child rearing life cycle, the study was limited to women between the ages of 20 and 50 who had at least one child of elementary school age living at home. Also, in order to avoid the initial adjustments associated with a new residential move, the study was restricted to women who had lived at least one year in their residence at the time of interview.

Women were interviewed who represented a 5 percent sample or approximately 100 of the households in each of the eight study neighborhoods.[94] The total sample equals 825 households or about 0.4 percent of the 195,000 households eligible[95] in the San Jose SMSA in 1975 - households, which in 1975, had a female head between the ages of 20-50, at least one child living at home, and an estimated 1975 family income of at least $10,000, a figure which represents the upper two-thirds of the 1975 family income distribution in the metropolitan area.

The American sample of this study was drawn from the following neighborhoods in the San Jose area in 1977:

| | American Neighborhood | Sample Size |
|---|---|---|
| A1 | Inlying single family neighborhood in San Jose | 105 |
| A2. | Outlying single family neighborhood in San Jose | 100 |
| A3. | Outlying planned unit developments in San Jose | 97 |
| A4. | Outlying condominiums in San Jose | 101 |
| A5. | Single family neighborhood in Los Gatos | 105 |
| A6. | Condominiums in Los Gatos | 103 |

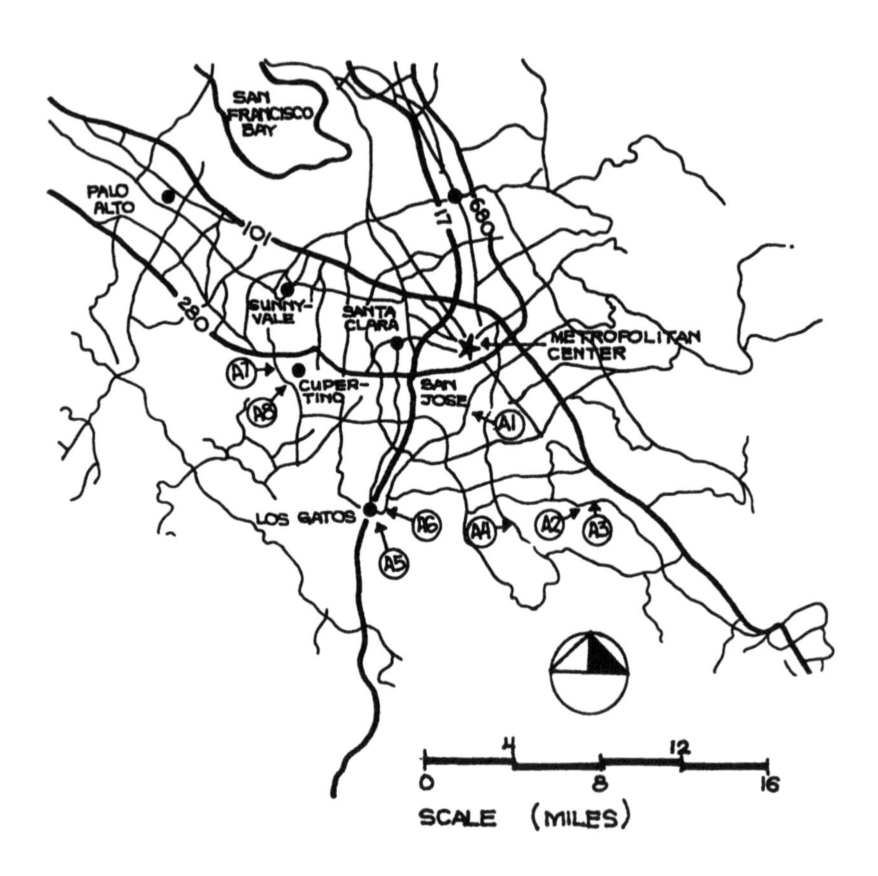


A1. INLYING S.F. (SAN JOSE)
A2. OUTLYING S.F. (SAN JOSE)
A3. OUTLYING PUD (SAN JOSE)
A4. OUTLYING CON (SAN JOSE)
A5. S.F. (LOS GATOS)
A6. CON (LOS GATOS)
A7. S.F. (CUPERTINO)
A8. CON (CUPERTINO)

FIGURE 3.4 LOCATION OF STUDY NEIGHBORHOODS IN THE SAN JOSE AREA, U.S.A.

| | | |
|---|---|---|
| A7. | Single family neighborhood in Cupertino | 109 |
| A8. | Condominiums in Cupertino | 105 |
| | Total | 825 |

As shown above and in Figure 3.4, the sample was spatially distributed among four residential environments in the central city of San Jose, (Neighborhoods A1, A2, A3, A4) and four residential areas in the nearby suburban communities of Los Gatos (Neighborhoods A5, A6) and Cupertino (Neighborhoods A7, A8). With the exception of Neighborhood A1 (Willow Glen), the inlying control neighborhood which is 2.0 miles from the center of San Jose, all neighborhoods are approximately 8.0 miles from the metropolitan center;[96] and all neighborhoods have similar economic and social characteristics.[97]

Table 3.1 reveals that with the possible exception of housing type (percent detached single-family)[98] the 1977 study samples of the central city and adjacent suburbs appear to be similar to the 1975 characteristics of middle-income families in those areas with respect to income, age of woman, marital status, size of household, and other household characteristics.[99] With regard to the total sample, Table 3.2 indicates that the study sample closely approximates the eligible households in the San Jose metropolitan area,[100] and is generally similar to the mean 1976 characteristics of comparable households in the suburban rings of all the SMSAs in the United States.[101] Thus, it appears that the American study households have qualities which generally typify much of the middle and upper-middle income suburban environments in the San Jose metropolitan area and in the United States.

2. Israel and the Netherlands

Employing similar techniques to those used in the San Jose study, socially matched middle class neighborhoods were selected for study in Israeli and Dutch metropolitan areas: three in the Tel-Aviv region; two in Jerusalem; and four in The Hague metropolitan area. As Figures 3.5 and 3.6 delineate, the two inlying Israeli neighborhoods (Bavly in Tel-Aviv, and St. Simone in Jerusalem) are about 2.0 miles from the central city center, while the three outlying areas (the new settlement Gilo in Jerusalem, and the suburban cities of Raanana and Kefar Sava in the Tel-Aviv area) range from 4.0 to 9.0 miles from the metropolitan center. The following is a breakdown of the Israeli sample by

TABLE 3.1 COMPARISON OF HOUSEHOLD CHARACTERISTICS OF STUDY SAMPLE OF CENTRAL CITY AND SUBURBAN RING IN THE SAN JOSE METROPOLITAN AREA WITH HOUSEHOLDS ELIGIBLE FOR STUDY IN THOSE AREAS

| | CENTRAL CITY | | SUBURBAN RING | |
|---|---|---|---|---|
| HOUSEHOLD CHARACTERISTICS | 1977 STUDY SAMPLE (n=403) | 1975 HOUSEHOLDS ELIGIBLE FOR STUDY IN THE SAN JOSE SMSA[b] (n=79,000) | 1977 STUDY SAMPLE (n=422) | 1975 HOUSEHOLDS ELIGIBLE FOR STUDY IN THE SAN JOSE SMSA[b] (n=116,000) |
| Family Income | 24,200[a] | 21,400[c] | 24,600[a] | 24,200[c] |
| Age of Woman | 33.7 | 32.9 | 35.6 | 34.1 |
| Marital Status (% married) | 90.1 | 87.3 | 80.7 | 79.8 |
| % Caucasian | 86.1 | 72.4[d]<br>88.9[f] | 92.3 | 80.8[e]<br>92.3[f] |
| Persons per Household | 4.06 | 4.21 | 3.87 | 4.01 |
| Monthly cost of household | 282 | 271 | 303 | 277 |
| Housing tenure (% owner occupied) | 93.5 | 85.7 | 88.3 | 88.6 |
| Housing Type (% detached single-family) | 75.0 | 62.1[d]<br>68.9[f] | 50.5 | 56.0[e]<br>70.7[f] |

a. 1976 income in U.S. Dollars reported during field interviews held in 1977

b. Households which had in 1975 a female head between the ages of 20-50; had at least one child of elementary school age living at home; and had an estimated 1975 annual income of at least $10,000

c. Estimated 1975 mean household income from extrapolating 1965-74 income trends for households having an estimated 1975 income of $10,000 or more

d. Mean value for the entire household population in the central city of the San Jose SMSA

e. Mean value for the entire household population in the suburban ring of the San Jose SMSA

f. 1975 mean value of census tracts from which study sample was drawn

Source: Santa Clara County Planning Department, _1975 Countywide Census, Santa Clara County_ (San Jose, 1976)

TABLE 3.2 COMPARISON OF HOUSEHOLD CHARACTERISTICS OF STUDY SAMPLE WITH HOUSEHOLDS ELIGIBLE FOR STUDY IN THE SAN JOSE METROPOLITAN AREA AND IN SUBURBAN RINGS OF ALL STANDARD METROPOLITAN STATISTICAL AREAS IN THE UNITED STATES

| HOUSEHOLD CHARACTERISTICS | 1977 STUDY SAMPLE (n=825) | 1975 HOUSEHOLDS ELIGIBLE FOR STUDY IN THE SAN JOSE SMSA[b.] (n=195,000) | 1976 COMPARABLE SUBURBAN HOUSEHOLDS IN ALL SMSAs[f.] (n=12,700,000) |
|---|---|---|---|
| Family Income | 24,400[a.] | 22,500[c.] | 21,000[g.] |
| Age of Woman | 34.7 | 33.5 | 34.2 |
| Marital Status (% married) | 85.4 | 83.5 | 80.5 |
| % Caucasian | 89.2 | 79.2[d.] 90.6[e.] | 91.6 |
| Persons per Household | 3.98 | 4.11 | 4.31 |
| Monthly Cost per Household | 290 | 273 | 256[h.] |
| Housing tenure (% owner occupied) | 90.1 | 86.2 | 75.7 |
| Housing type (% detached single-family) | 63.0 | 61.3[d.] 69.8[e.] | 72.3[h.] |

a. 1976 income reported during field interviews held in 1977
b. Households which had in 1975 a female head between the ages 20-50; had at least one child of elementary school age living at home; and had an estimated 1975 annual income of at least $10,000
c. Estimated 1975 mean household income from extrapolating 1965-74 income trends for households having an estimated 1975 income of $10,000 or more
d. Mean value for the entire household population for the San Jose SMSA
e. 1975 mean value of census tracts from which study sample was drawn.
f. Households which were in 1976 in the suburban rings of all SMSAs (as defined in 1970) and had at least one child less than 18 years old living at home
g. 1975 mean income for owner occupied households
h. 1975 mean value from a national survey of 23 typical SMSAs

TABLE 3.2 (Continued)

Sources: All San Jose SMSA data from Santa Clara County Department, 1975 Countywide Census, Santa Clara County (San Jose, 1976)
Data for suburban rings of all U.S. SMSAs from: U.S. Department of Commerce, Bureau of the Census, Consumer Income: Household Money Income in 1975 By Housing Tenure and Residence for the United States, Regions, Divisions and States (U.S. Government Printing Office, Washington D.C., 1977); except 'Age of Woman' data from U.S. Department of Commerce, Bureau of the Census, Household and Family Characteristics (U.S. Government Printing Office, Washington D.C., 1977); and except 'Monthly Cost of Household' and 'Housing Type' data from U.S. Department of Commerce, Bureau of the Census Annual Housing Survey: 1975 Summary of Housing Characteristics for Selected Metropolitan Areas (U.S. Government Printing Office, Washington D.C., 1978)

neighborhood in 1978:

| | Israeli Neighborhood | Sample Size |
|---|---|---|
| B1. | Inlying less planned Bavly neighborhood in Tel-Aviv | 50 |
| B2. | Outlying less planned areas in Raanana and Kefar Sava | 65 |
| B3. | Outlying planned areas in Raanana and Kefar Sava | 60 |
| B4. | Inlying planned St. Simone neighborhood in Jerusalem | 60 |
| B5. | Outlying planned neighborhood in Gilo town in Jerusalem area | 60 |
| | Total | 295 |

In The Hague, Figure 3.7 shows that two inlying neighborhoods (Old and New Leyenburg) are approximately 2.0 miles from the city center and the two outlying areas (Bouwlust in The Hague and TeWerve West in the adjoining suburban city of Rijswijk) are about 4.0 miles from the center. The Dutch sample of this study was taken from the following areas in The Hague region in 1978:

| | Dutch Neighborhood | Sample Size |
|---|---|---|
| C1. | Inlying less planned Old Leyenburg neighborhood | 49 |
| C2. | Inlying planned New Leyenburg neighborhood | 56 |

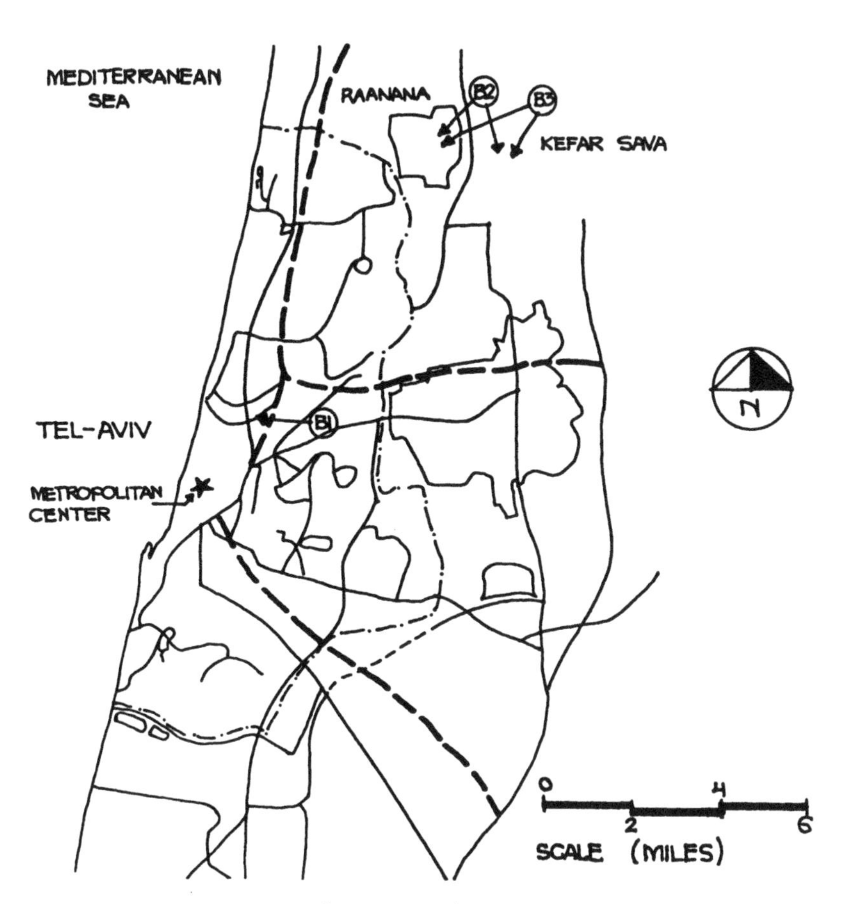


B1. INLYING BAVLY (TEL-AVIV)
B2. OUTLYING LESS PLANNED (RAANANA AND KEFAR SAVA)
B3. OUTLYING PLANNED (RAANANA AND KEFAR SAVA)

FIGURE 3.5 LOCATION OF STUDY NEIGHBORHOODS IN THE TEL-AVIV AREA, ISRAEL

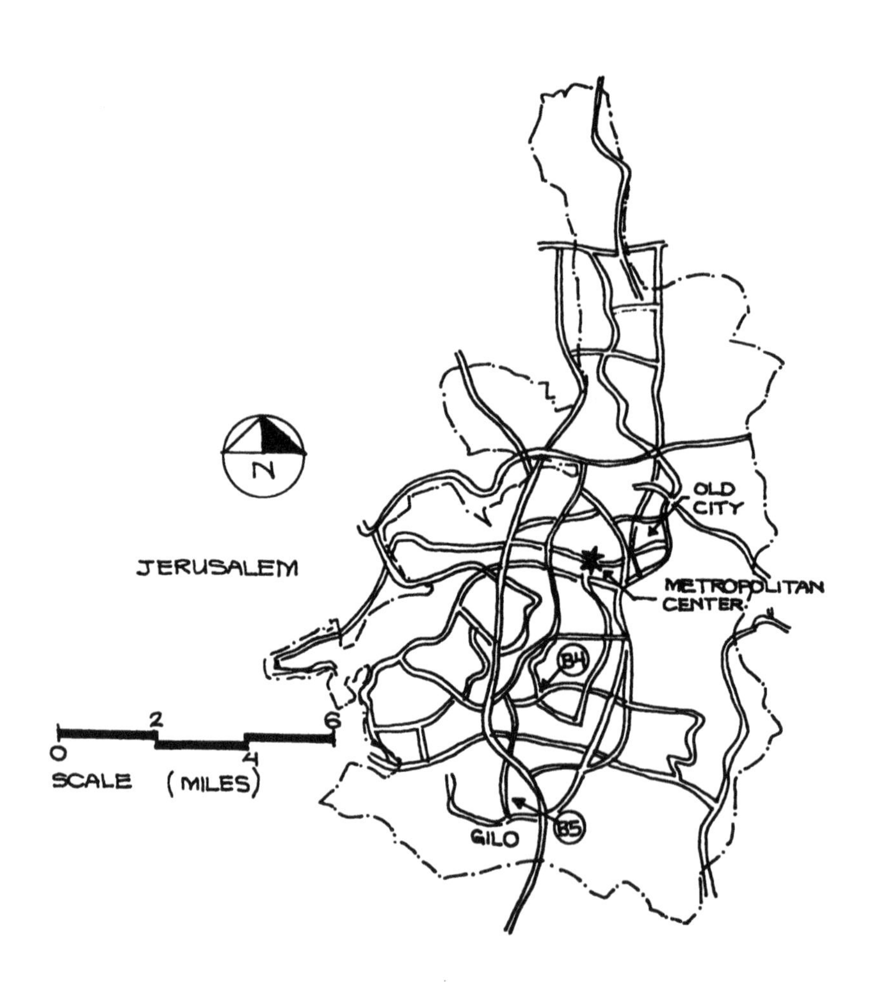


B4. INLYING ST. SIMONE AREA
B5. OUTLYING GILO TOWN

FIGURE 3.6 LOCATION OF STUDY NEIGHBORHOODS IN THE JERUSALEM AREA, ISRAEL

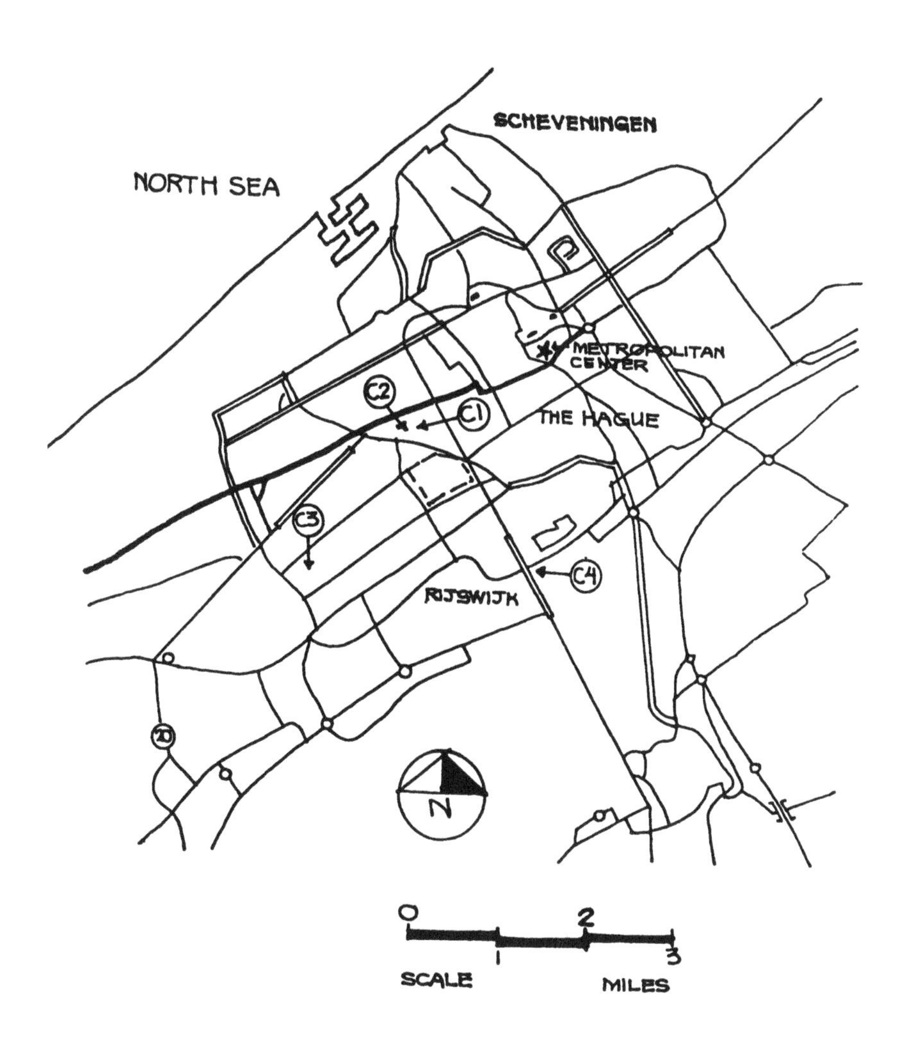


C1. INLYING OLD LEYENBURG (THE HAGUE)
C2. INLYING NEW LEYENBURG (THE HAGUE)
C3. OUTLYING BOUWLUST (THE HAGUE)
C4. OUTLYING SUBURB (RIJSWIJK)

FIGURE 3.7 LOCATION OF STUDY NEIGHBORHOODS IN THE HAGUE AREA, THE NETHERLANDS

| | | |
|---|---|---|
| C3. | Outlying planned Bouwlust area of The Hague | 57 |
| C4. | Outlying less-planned Tewerve West in Rijswijk | 53 |
| | Total | 215 |

The sample of female respondents in the Israeli and Dutch study neighborhoods was chosen randomly, with the same age, child and length of residency requirements as those of the American study. The women interviewed (295 in Israel; and 215 in the Netherlands) represented about 4 percent of their respective neighborhood households, and came from typical middle income households in metropolitan areas of their respective countries (see Tables 3.3 and 3.4)

## C. DATA COLLECTION

Data were collected through the use of a structured questionnaire administered by an interviewer in the home of each subject. Each personal interview was conducted by a professionally trained female researcher and lasted about one hour.

## D. DATA ANALYSIS

After the data were collected, scores from questions concerning each variable were tabulated and transferred to computer cards for statistical analysis. First, simple correlation coefficients were computed in order to make a preliminary examination of the relationships between variables. This procedure also acted as a technique for screening out independent variables with marginal influences on dependent variables. Other statistical techniques, such as t-tests and analysis of variance were used to test differences between mean scores of sub-groups of the study sample (e.g., multi-family vs. single-family residents). Partial correlation coefficients were employed primarily to test for expected relationships between independent and dependent variables in the entire sample.[102] In some cases, multiple regression analysis was used to test the combined impact of the independent variables expected to influence each dependent variable.

Because of the exploratory nature of this study, only tentative expectations of the relationships among the variables examined were used and thus two-tailed t-tests of significance seem

TABLE 3.3 COMPARISON OF HOUSEHOLD CHARACTERISTICS OF ISRAELI STUDY SAMPLE WITH HOUSEHOLDS IN URBAN AREAS AND TOTAL NATION OF ISRAEL

| HOUSEHOLD CHARACTERISTICS | 1978 STUDY SAMPLE[b] (n=295) | 1978 URBAN HOUSEHOLDS[c] (n=868,300) | 1978 TOTAL HOUSEHOLDS IN ISRAEL (n=974,400) |
|---|---|---|---|
| Family Income[a] | 98,900 | 105,300 | 98,100 |
| Age of Woman | 34.8 | 35.7 | 32.3[e] |
| Marital Status (% married) | 96.4 | 95.8 | 95.1 |
| Percent Sephardim[d] | 21.7 | 33.4[f] | 27.9[e] |
| Persons per Household | 4.32 | 4.34[g] | 4.52[g] |
| Monthly cost of household | 2,240 | 2,480 | 2,180 |
| Housing tenure (% owner occupied) | 91.2 | 70.4 | 70.6 |
| Housing type (% detached single family) | 0 | 12.3 | 13.1 |

a. 1978 income in Israeli Pounds reported during field interviews held in May 1978
b. Data reported during our field interviews held in 1978
c. Localities whose population numbered 10,000 inhabitants or more
d. Jews born in Asian or African countries
e. Percent women between the ages of 20-50
f. Percent households with head of household born in Asia or Africa
g. Households which had in 1978 at least one child up to 17 years old living at home

Sources: Israel, Central Bureau of Statistics, _Survey of Housing Conditions: 1978_ (Jerusalem, 1980); Israel, Central Bureau of Statistics, _Statistical Abstract of Israel: 1978, 1979, 1980_ (Jerusalem, 1979, 1980, 1981)

TABLE 3.4 COMPARISON OF HOUSEHOLD CHARACTERISTICS OF DUTCH STUDY SAMPLE WITH HOUSEHOLDS IN THE HAGUE METROPOLITAN AREA AND TOTAL NATION OF THE NETHERLANDS

| | 1978 HOUSEHOLDS IN THE HAGUE AREA | | | |
|---|---|---|---|---|
| | 1978[b.] STUDY SAMPLE (n=215) | CENTRAL CITY (n=203,800) | SUBURBAN RING[c.] (n=75,830) | 1978 TOTAL HOUSEHOLDS IN THE NETHERLANDS (n=4,768,000) |
| Family Income[a.] | 27,100 | 24,770[d.] | 30,470[d.] | 25,180[d.] |
| Age of Head of Household | 37.6 | 52.0 | 49.0 | 48.0 |
| Marital Status of Head of Household (% married) | 96.3 | 50.0 | 67.0 | 67.0 |
| Persons per Household | 4.05 | 2.28 | 2.75 | 2.76 |
| Monthly cost of Housing | 253 | 220 | 280 | 203 |
| Housing tenure (% owner occupied) | 58.2 | 28.0 | 34.0 | 42.0 |
| Housing type (% detached single family) | 0 | 2.0 | 5.0 | 19.0 |

a. 1978 income in Dutch Guilders reported during field interviews held in 1978
b. Data reported during our field interviews held in 1978
c. i.e., Leidschendam, Nootdorp, Rijswijk, Voorburg, Wassenaar, Zoetermeer
d. Estimates deflated from 1981 data in The Netherlands, Housing Demand Survey

Sources: The Netherlands, Housing Demand Survey, October 1981); The Netherlands, Central Bureau of Statistics, Statistical Yearbook in the Netherlands: 1980 (Staatsuitgeverj, The Hague, 1981)

most appropriate. Of course, it is recognized in the social sciences that it is desirable to obtain at least a 95 percent probability of no error due to chance ($p<0.05$) before granting any theoretical importance to the relationships uncovered. However, in order to call the attention of the reader to potentially important areas for future research, results are reported with a somewhat lower 90 percent probability of no error due to chance ($p<0.10$).

NOTES

[1] John B. Lansing, et al., Planned Residential Environments (University of Michigan Press, Ann Arbor, 1979).

[2] James R. Anderson, et al., 'Resident's Satisfaction: Criteria for the Evaluation of Housing for Low and Moderate Income Families', paper delivered at the American Insitute of Planners Conference, Denver, October 1974.

[3] Robert B. Zehner, Indicators of the Quality of Life in the New Communities (Ballinger Publishing, Cambridge, Mass., 1977).

[4] Claude S. Fischer, To Dwell Among Friends: Personal Networks in Town and City (University of Chicago Press, Chicago, 1982).

[5] Angus Campbell, Philip E. Converse and Willard L. Rodgers, The Quality of American Life (Russell Sage Foundation, New York, 1976); Environmental Protection Agency, The Quality of Life Concept (Environmental Protection Agency, Washington, D.C., 1973); Henry J. Schmandt and Warner Bloomberg, Jr., (eds.), The Quality of Urban Life (Sage Publications, Beverly Hills, 1969).

[6] William R. Ewald, Jr., (ed.), Environment for Man (Indiana University Press, Bloomington, 1967), pp.3-109; Claude S. Fischer and Robert Max Jackson, 'Suburbs, Networks and Attitudes' in Barry Schwartz (ed.), The Changing Face of the Suburbs (University of Chicago Press, Chicago, 1976), pp.279-307; Charles Haar (ed.), The President's Task Force on Suburban Problems (Ballinger Publishing Company, Cambridge, Mass., 1974); Louis Harris Associates, A Survey of Citizen Views and Concerns About Urban Life (Department of Housing and Urban Development, Washington, D.C., 1978); Fischer, To Dwell Among Friends; William Michelson, Environmental Choice, Human Behavior and Residential Satisfaction (Oxford University Press, New York, 1977); Harvey S. Perloff, 'A Framework for Dealing

with the Urban Environment: An Introductory Statement', in The Quality of the Urban Environment (Johns Hopkins Press, Baltimore, 1969), pp.3-31; Schmandt and Bloomberg, Quality of Urban Life; Nestor E. Terleckyi, Improvements in the Quality of Life: Estimates of Possibilities in the United States (National Planning Association, Washington, D.C., 1975); Zehner, Quality of Life.

[7] Alden E. Wessman and David F. Ricks, Mood and Personality (Holt, Rinehart and Winston, New York, 1966).

[8] Campbell, Converse and Rodgers, Quality of American Life.

[9] Zehner, Quality of Life.

[10] Wessman and Ricks, Mood and Personality.

[11] Ibid.

[12] Campbell, Converse and Rodgers, Quality of American Life.

[13] See for example, Mark Baldassare and Claude S. Fischer, 'Suburban Life: Powerlessness and Need for Affiliation', Urban Affairs Quarterly, vol.10, no.3 (1975), pp.314-26; Herbert J. Gans, The Levittowners, (Vantage Books, New York, 1967); Harvey Marshall, 'Suburban Life Styles: A Contribution to the Debate', in Louis H. Masotti and Jeffrey K. Hadden (eds.), The Urbanization of the Suburbs (Sage Publications, Beverly Hills, 1973), pp.123-48.

[14] See: Sherman Maisel and Louis Winnick, 'Family Housing Expenditures: Illusive Laws and Intrusive Variances', in William L. C. Wheaton, et al., (eds.), Urban Housing (Free Press, New York, 1966), pp.139-53; Israel, Central Bureau of Statistics, Statistical Abstract of Israel: 1978 (Jerusalem, 1978); The Netherlands, Social and Cultural Planning Office, Social and Cultural Report: 1978 (Rijswijk, 1978).

[15] Campbell, Converse and Rodgers, Quality of American Life; Gans, The Levittowners: Thomas A. C. Rennie, et al., Mental Health in the Metropolis (McGraw-Hill, New York, 1962); William Rushing, 'Two Patterns in the Relationship Between Social Class and Mental Hospitalization', American Sociological Review, vol.34, no.4 (1969), pp.533-41.

[16] Arnold Feldman and Charles Tilly, 'The Interaction of Social and Physical Space', American Sociological Review, vol.25, no.6 (1969), pp.877-84.

[17] Lansing, et al., Planned Environments .

[18] Wendell Bell and Marian Boat, 'Urban Neighborhoods and Informal Social Relations', American Journal of Sociology, vol.62, no.3 (1957),

pp.391-8; William Michelson, 'Environmental Change', Centre for Urban and Community Studies, Research Paper No. 60, October, 1973.

[19] For example, see G. A. Allan, A Sociology of Friendship and Kinship (Allen and Unwin, London, 1979).

[20] Campbell, Converse and Rodgers, Quality of American Life.

[21] William Michelson, Man and His Urban Environment (Addison-Wesley, Reading, Mass., 1976).

[22] Cynthia Epstein, Woman's Place (University of California Press, Berkeley, 1970); Juanita Kreps, Sex in the Market Place (Johns Hopkins Press, Baltimore, 1971).

[23] Walter R. Gove and Jeannette F. Tudor, 'Adult Sex Roles and Mental Illness', in Joan Huber (ed.), Changing Women in a Changing Society (University of Chicago Press, Chicago, 1973), pp.50-73.

[24] Campbell, Converse and Rodgers, Quality of American Life; James D. Wright, 'Are Working Women Really More Satisfied?' Journal of Marriage and the Family, vol.40, no.2 (1978), pp.301-14.

[25] Herbert J. Gans, The Urban Villagers (Free Press of Glencoe, New York, 1962); Gans, The Levittowners.

[26] Michelson, Man and His Urban Environment.

[27] Gans, The Levittowners; Michelson, Environmental Choice.

[28] For the United States see Nathan Glazer and Daniel P. Moynihan, Beyond the Melting Pot (MIT Press, Cambridge, Mass., 1963); Nathan Kantrowivz, 'Ethnic and Racial Segregation in the New York Metropolis, 1960', American Journal of Sociology, vol.74, no.6 (1969), pp.685-95; and Michael Novak, 'How American Are You If Your Grandparents Came from Serbia in 1888?' in Sallie Te Selle (ed.), The Rediscovery of Ethnicity: Its Implications for Culture and Politics in America (Harper and Row, New York, 1973), pp.1-20. For Israel see Vivian Klaff, 'Residence and Integration in Israel: A Mosaic of Segregated Groups', and Sammy Smooha and Yachanan Peres, 'The Dynamics of Ethnic Inequalities: The Case of Israel' in Ernest Krauz (ed.), Studies of Israeli Society, vol.I, (Transaction Books, London, 1980), pp.53-71 and 165-181. For the Netherlands, see Francois J. Gay, 'Benelux', in Hugh D. Clout, (ed.), Regional Development in Western Europe (John Wiley, New York, 1981), pp.179-209; and Dieter Lappk and Pieter van Hoogtraten, 'Remarks on the Spatial Structure of Capitalist Development: the

Case of the Netherlands', in John Carney, Ray Hudson and Jim Lewis (eds.), Regions in Crisis: New Perspectives in European Regional Theory (St. Martins Press, 1980), pp.117-71.

[29] For inter-regional migration in Europe, see Hugh D. Clout, 'Population and Urban Growth', in Hugh D. Clout (ed.), Regional Development in Western Europe, pp.35-9. For immigration to Israel, see Haim Darin-Drabkin, Housing In Israel: Economic and Sociological Aspects (Gadish Books, Tel-Aviv, 1957), pp.19-51.

[30] Derk de Jonge, 'Some Notes on Sociological Research in the Field of Housing', Mimeo, Delft University of Technology, 1967.

[31] Claude S. Fischer, 'Toward a Subcultural Theory of Urbanism', American Journal of Sociology, vol.80, no.6 (1975), pp.1319-41; de Jonge, 'Some Notes on Sociological Research'.

[32] Gans, The Levittowners.

[33] Michelson, Environmental Choice.

[34] Gans, The Levittowners; Clark E. Moustakas, Loneliness and Love (Prentice-Hall, Englewood Cliffs, 1972); Robert S. Weiss, et al. Loneliness (MIT Press, Cambridge, Mass., 1973).

[35] Norman Schulman, 'Mutual Aid and Neighboring Patterns: The Lower Town Study', Anthropoligica, vol.9, no.1 (1967), pp.51-60; E. Pfeil, 'The Pattern of Neighboring Relations in Dortmund - Nordstadt', in R. E. Pahl (ed.), Readings in Urban Sociology (Pergamon Press, London, 1968), pp.136-58.

[36] Gans, The Levittowners.

[37] Baldassare and Fischer, 'Suburban Life'; Michelson, Environmental Choice.

[38] Baldassare and Fischer, 'Suburban Life'; Michelson, Environmental Choice.

[39] J. Douglas Porteous, Environment and Behavior: Planning and Everyday Urban Life (Addison-Wesley Publishing Company, Reading, Mass., 1977).

[40] S. D. Clark, The Suburban Society (University of Toronto Press, Toronto, 1966).

[41] Gans, The Levittowners.

[42] Karen Hapgood and Judith Getzels, (eds.), Planning, Women and Change (American Society of Planning Officials, 1974), pp.1-32; Campbell, Converse and Rodgers, Quality of American Life.

[43] See, for example Robert N. Butler and Myrna I. Lewis, Aging and Mental Health (C. V. Mosby, St. Louis, 1977); Rennie, et al., Mental Health in the Metropolis; and Fischer, To Dwell Among Friends, p.55.

[44] Elizabeth M. Havens, 'Women, Work and Wedlock: A Note on Female Marital Patterns in the United States', in Joan Huber, (ed.), Changing Women in a Changing Society (University of Chicago Press, Chicago, 1973), pp.212-9.

[45] U.S. Bureau of the Census, 'Household Income in 1972 and Selected Social and Economic Characteristics of Households Series', p.60-80, (1972); Campbell, Converse and Rodgers, Quality of American Life.

[46] Hapgood and Getzels, Planning, Women and Change.

[47] See Economic and Social Opportunities, Inc., Female Heads of Household and Poverty in Santa Clara County (Economic and Social Opportunities, San Jose, 1974); Fischer, To Dwell Among Friends, p.55; and Bernard Rosenberg, 'Women's Place in Israel', Dissent, vol.24, no.4 (Fall, 1977), pp.408-17.

[48] Marilyn M. Pray, 'Planning and Women in the Suburban Setting', in Hapgood and Getzels, Planning, Women and Change, pp.51-6.

[49] Scott Greer, 'The Family in Suburbia', in Louis H. Masotti and Jeffrey K. Hadden, (eds.), The Urbanization of the Suburbs, (Sage Publications, Beverly Hills, 1973), pp.149-70.

[50] Michelson, Man and His Urban Environment.

[51] Ibid.

[52] Melanie Freitas, 'Woman in Suburbia', unpublished Master's Planning Report, San Jose State University, 1974.

[53] Hapgood and Getzels, Planning, Women and Change; Economic and Social Opportunities, Inc., Female Heads of Household; Campbell, Converse and Rodgers, Quality of American Life, pp.406-9.

[54] See, for example, Tommy Garling, Anders Book, and Erik Lindsberg, 'Cognitive Mapping of Large-Scale Environments: The Interrelationship of Action Plans, Acquisition, and Orientation', Environment and Behavior, vol.16, no.1 (1984), pp.3-34; Michelson, Man and His Urban Environment; Gary T. Moore, 'Knowing About Environmental Knowing: The Current State of Theory and Research on Environmental Cognition', Environment and Behavior, vol.11, no.2 (1979), pp.33-70; Humphrey Osmund, 'Some Psychiatric Aspects of Design', in Laurence B. Holland (ed.), Who Designs America? (Anchor Books, Garden City, 1966), pp.281-318; Donald N. Rothblatt, 'Improving the Design of Urban Housing', in Vasily Kouskoulas (ed.), Urban Housing (National Sciences Foundation, Detroit, 1973), pp.149-54; Henry Sanoff, 'Neighborhood Satisfaction: A Study of

User Assessments of Low Income Residential Environment', in Oktal Ural (ed.), Proceedings of the Second International Symposium on Lower-Cost Housing Problems, (University of Missouri-Rolla, St. Louis, 1973), pp.119-24; and Zehner, Quality of Life.

[55] See Gans, The Levittowners: Nathan Glazer, 'Slum Dwellings Do Not Make A Slum', New York Times Magazine (November 21, 1965), p.55; and Daniel M. Wilner, et al., The Housing Environment and Family Life (Johns Hopkins Press, Baltimore, 1962); and Zehner, Quality of Life.

[56] Lansing, et al., Planned Environments.

[57] Myrna M. Weissman and Eugene S. Paykel, 'Moving and Depression in Women', in Robert S. Weiss (ed.), Loneliness, (MIT Press, Cambridge, 1973), pp.154-64.

[58] Michelson, Man and His Urban Environment.

[59] Donald N. Rothblatt, 'Housing and Human Needs', Town Planning Review, vol.42, no.2, (1971), pp.130-44; and Aida K. Tomeh, 'Informal Group Participation and Residential Patterns', American Journal of Sociology, vol.70, no.1 (1964), pp.28-35.

[60] Michelson, Man and His Urban Environment.

[61] Scott Greer, The Urbane View: Life and Politics in Metropolitan America (Oxford University Press, New York, 1972); and Eshref Shevky and Wendell Bell, Social Area Analysis (Stanford University Press, Stanford, 1955).

[62] John Raven, 'Sociological Evidence on Housing (2: The Home Environment)', The Architectural Review, vol.142, no.1 (1967), pp.236ff.

[63] Michelson, Environmental Choice.

[64] See John F. Kain, 'The Distribution and Movement of Jobs and Industry', in James Q. Wilson (ed.), The Metropolitan Enigma (United States Chamber of Commerce, Washington, 1967), pp.1-31; and Clout, 'Population and Urban Growth'.

[65] Pray, 'Planning and Women'.

[66] Phyllis Kaniss and Barbara Robins, 'The Transportation Needs of Women', in Hapgood and Getzels, Planning, Women and Change, pp.63-70.

[67] John R. Meyer, 'Urban Transportation', in James Q. Wilson (ed.), The Metropolitan Enigma, pp.34-75.

[68] Michelson, Environmental Choice; Porteous, Environment and Behavior.

[69] Clark, Suburban Society.

[70] Baldassare and Fischer, 'Suburban Life'; Michelson, Environmental Choice.

[71] See, for example, John R. Meyer, John F. Kain and Martin Wohl, The Urban Transportation

*Problem* (Harvard University Press, Cambridge, Mass., 1965); and Mark La Gory and John Pipkin, *Urban Social Space* (Wadsworth Publishing Co., Belmont, Cal., 1981), Chapter 7.

[72] John F. Kain, 'The Journey to Work as a Determinant of Residential Location', in Alfred N. Page and Warren R. Segfried (eds.), *Urban Analysis* (Scott, Foresman, Glenview, Ill., 1970), pp.207-26.

[73] William Alonso, 'A Theory of the Urban Land Market', in Matthew Edel and Jerome Rothenberg (eds.), *Readings in Urban Economics* (MacMillan, New York, 1972), pp.104-11; La Gory and Pipkin, *Urban Social Space*, Chapter 7.

[74] Kain, 'Journey to Work'.

[75] See Howard S. Lapin, *Structuring the Journey to Work* (University of Pennsylvania Press, Philadelphia, 1964); John Wolforth, 'The Journey to Work', in Larry S. Bourne (ed.), *Internal Structure of the City* (Oxford University Press, New York, 1971), pp.240-7; and Marshall, 'Suburban Life Styles'.

[76] Julie A. Erickson, 'An Analysis of the Journey to Work for Women', *Social Problems*, vol.24, no.4 (1977), pp.428-35.

[77] Clark, *Suburban Society*.

[78] Gans, *The Levittowners*; Baldassare and Fischer, 'Suburban Life'.

[79] For example, see Robert C. Wood, *1400 Governments* (Harvard University Press, Cambridge, Mass., 1961); Robert L. Lineberry and Ira Sharksansky, *Urban Politics and Public Policy* (Harper and Row, New York, 1978); Amiram Gonen, 'The Suburban Mosaic in Israel', in D. H. K. Amiran and Y. Ben-Arieh (eds.), *Geography in Israel* (Israel National Committee, Jerusalem, 1976), pp.163-86; and John R. Short, 'Urban Policy and British Cities', *Journal of the American Planning Association*, vol.48, no.1 (1982), pp.39-52.

[80] Oliver P. Williams, *Metropolitan Political Analysis* (Free Press, New York, 1971).

[81] Robert L. Bish, *The Public Economy of Metropolitan Areas* (Markham Publishing, Chicago, 1971); Henry J. Schmandt and G. Stephens, 'Measuring Municipal Output', *National Tax Journal*, vol.13, no.4 (1960), pp.369-75.

[82] Robert C. Wood, *Suburbia: Its People and Their Politics* (Houghton Mifflin, Boston, 1958).

[83] Louis H. Masotti and D. Bowen, 'Communities and Budgets: The Sociology of Municipal Expenditures', *Urban Affairs Quarterly*, vol.1, no.1 (1965), pp.38-58; Bryan T. Downes, 'Suburban

Differentiation and Municipal Policy Choices', in Terry N. Clark (ed.), Community Structure and Decision-Making (Chandler Publishing Company, San Francisco, 1968), pp.243-67.

[84] Robert L. Bish and Hugh O. Nourse, Urban Economics and Policy Analysis (McGraw-Hill, New York, 1975); C. M. Tiebout, 'A Pure Theory of Local Expenditures', in Matthew Edel and Jerome Rothenberg (eds.), Readings in Urban Economics, (MacMillan, New York, 1972), pp.513-23.

[85] See for example, Edward P. Eichler and Marshall Kaplan, The Community Builders (University of California Press, Berkeley, 1967); William H. Whyte, The Last Landscape (Doubleday, New York, 1969); David R. Goldfield, 'National Urban Policy in Sweden', Journal of the American Planning Association, vol.48, no.1 (1982), pp.24-38; and Gonen, 'The Suburban Mosaic in Israel'.

[86] Joseph Zelan, 'Does Suburbia Make a Difference?' in Sylvia F. Fava (ed.), Urbanism in World Perspective (Crowell, New York, 1968), pp.401-8 .

[87] Sylvia F. Fava, 'Beyond Suburbia', Annals of the American Academy of Political and Social Science, vol.422 (1957), pp.10-24.

[88] Michelson, Man and His Urban Environment; Carl Werthman, et al., Planning and the Purchase Decision: Why People Buy in Planned Communities (University of California Press, Berkeley, 1965).

[89] Lansing, et al., Planned Environments; Michelson, Environmental Choice .

[90] Francine Rabinovitz and James Lamare, 'After Suburbia, What?' in Werner Z. Hirsch, (ed.), Los Angeles: Viability and Prospects for Metropolitan Leadership, (Praeger, New York 1971), pp.169-206.

[91] For example, see the evaluation of Vallingby, the planned suburb in the Stockholm metropolitan area, in David Poponoe, The Suburban Environment: Sweden and the United States (University of Chicago Press, Chicago, 1977).

[92] This includes a variety of planned environments, such as planned unit developments, condominium complexes and other residential areas having special design and site plan characteristics (e.g., community facilities integrated into the overall site plan).

[93] See Santa Clara County Planning Department, 1975 Countywide Census, Santa Clara County (1976).

[94] The mean number of occupied households in each census tract in the San Jose SMSA was 1,870. Ibid.

[95] These 195,000 households or approximately 50% of the total of 392,400 households in the San Jose SMSA were found eligible for our study using the criteria of income, age of female head and children.

[96] For neighborhoods A2 through A7, the mean household distance from the central city center ranged from 7.7 to 8.5 miles.

[97] For all neighborhoods, the mean 1975 family income varied from $22,620 to $26,580, the mean level of education was at least high school graduation.

[98] While the Housing Type (percent detached single family) was considerably lower in our suburban sample than in the actual suburban population, it was virtually unavoidable. In order to have our sample approximate the percentage of multiple dwellings for total metropolitan area at the income, size and tenure levels of our research, we needed, of course, to conduct our interviews where appropriate multiple dwellings exist - often in the adjacent suburbs.

[99] In order to attain our objective of studying neighborhoods of equal population size, a greater proportion of the populations of Los Gatos (3.3%) and Cupertino (3.8%) are included in the total sample than those from San Jose (0.6%). To the extent that these two political units may have social, economic and physical characteristics not typical of the San Jose metropolitan area, care should, of course, be taken into account when generalizing about the population (see Table 3.1). That is, one-half of our sample comes from two areas which are a much smaller proportion of the total SMSA. Turning this analysis around, however, there is a general correspondence in a broader sense in that San Jose constitutes about one-half of the population of the SMSA and other small communities (of which Los Gatos and Cupertino are examples) constitute the other half.

[100] Although the percent caucasian for our sample study of 89.2 was substantially higher than the mean value for the total San Jose SMSA of 79.2, it was quite similar to the mean value of the middle and upper-middle income census tracts from which our sample was drawn of 90.6.

[101] Households in the suburban rings of all SMSAs with at least one child less than 18 years old living at home in 1976. Based on SMSA definitions of 1970.

[102] The independent variables used in our

partial correlation analysis are: income, education, occupational status, ethnic identity, extent of urban origins, distance of regional origin from metropolitan area, length of residency, age, number and age of children, age of neighborhood, distance to work, ratio of family income to mean neighborhood family income, and ratio of woman's education to mean neighborhood woman's education. Due to the relative openness of Israel and the Netherlands to international migration, the following additional variables were included in the partial correlation analysis for these countries: country of birth for woman and man, and length of national residency for woman and man. Because of the almost binary nature of the other independent variables, they were examined with t-tests or analysis of variance.

# CHAPTER 4

# FINDINGS AND DISCUSSION

In this chapter the findings and conclusions of the tri-national study are presented. These judgements describe the comparative influence of the four categories of independent variables - social class, subcultural, life cycle, and suburban environmental - on levels of satisfaction in four realms of the quality of life - housing environment, community services, social patterns, and psychological well-being. For each independent variable our expectations are stated, our findings are presented, and their implications are discussed.

## A. SOCIAL CLASS INFLUENCES

### 1. Income

Expectations: Income will vary positively with all dependent satisfaction variables.

Findings: The data provides partial support for this contention in the United States, strong support in Israel, and virtually no support in the Netherlands (see Table 4.1). As expected, there does exist a strong positive relationship between income and the overall index of satisfaction with the housing environment in Israel and in the United States ($p<0.010$ for both). Particularly striking within this measure is the strong linkage between income and satisfaction with household responsibilities ($p<0.005$ for both).

With respect to community services, no support was provided by our Dutch data, slight support by that from Israel, with rather mixed results from the United States. While no statistical relationship was generated from the Dutch sample, satisfaction with entertainment from the Israeli sample conformed

TABLE 4.1 RELATIONSHIPS BETWEEN FAMILY INCOME AND INDEXES OF SATISFACTION: U.S.A., ISRAEL AND THE NETHERLANDS

| Satisfaction Index | Partial Correlation Coefficient[a] | | |
|---|---|---|---|
| | U.S.A. (n=825) | Israel (n=295) | Netherlands (n=215) |
| a. Housing Environment | 0.0971[d] | 0.1735[d] | 0.0377 |
| 1. Privacy | -0.0004 | 0.0640 | 0.0039 |
| 2. House & Lot | 0.0243 | 0.0506 | 0.0326 |
| 3. Household Responsibilities | 0.1518[e] | 0.3508[e] | 0.0584 |
| 4. Appearance | 0.0559 | -0.0074 | 0.0106 |
| b. Community Services | 0.0186 | 0.0878 | 0.0373 |
| 1. Parks | -0.0868[c] | 0.0008 | -0.0139 |
| 2. Schools | 0.0093 | -0.0259 | -0.0759 |
| 3. Security | 0.1120[e] | 0.0350 | 0.1010 |
| 4. Child Care | -0.0303 | -0.0221 | 0.0275 |
| 5. Transportation | 0.1054[e] | -0.0148 | -0.0097 |
| 6. Entertainment | -0.0660[b] | 0.1973[e] | -0.0220 |
| c. Social Patterns | 0.0552 | 0.2618[e] | 0.1037 |
| 1. Friendships | 0.0422 | 0.1403 | 0.0669 |
| 2. Group Activities | 0.0488 | 0.3185[e] | 0.1896[c] |
| 3. Sense of Belonging | 0.0411 | -0.0351 | -0.0150 |
| d. Psychological Well-being | -0.0130 | 0.1000 | -0.0394 |
| 1. Fullness vs. Emptiness of Life | 0.0423 | 0.1203[b] | -0.0301 |
| 2. Social Respect | -0.0452 | -0.0370 | -0.0487 |
| 3. Personal Freedom | 0.0173 | 0.0474 | -0.0387 |
| 4. Companionship | -0.0330 | 0.0771 | 0.0500 |
| 5. Tranquility vs. Anxiety | 0.0524 | 0.1142[b] | -0.0336 |
| 6. Self Approving vs. Guilty | -0.0497 | 0.0048 | -0.0823 |
| 7. Self Confidence | -0.0357 | -0.0260 | -0.0310 |
| 8. Elation vs. Depression | 0.0014 | 0.1029 | 0.0081 |

a. two-tailed t-test yields $p>0.100$ unless otherwise noted
b. t-test yields $p<0.100$
c. t-test yields $p<0.050$
d. t-test yields $p<0.010$
e. t-test yields $p<0.005$

to expectations. American data showed no overall positive relationship between income and community services, but satisfaction with security and transportation lent strength to the hypothesis (both $p<0.005$). However, two additional components, parks ($p<0.050$) and entertainment ($p<0.100$) both demonstrated a negative relationship with income, contrary to expectations.

While satisfaction with the social patterns cluster of dependent variables failed to support expectations in the American sample, very strong support was present in data generated in Israel, and some support from that in the Netherlands. Not only was overall satisfaction with social patterns a very strong correlate of income in Israel ($p<0.005$), but so were satisfaction with group activities ($p<0.005$) and with friendships ($p<0.050$). Satisfaction with group activities ($p<0.050$) in the Netherlands echoed the Israeli findings.

Measures of psychological well-being failed to show a statistical relationship to income in all three countries, with the exception of two measures in Israel which fell short of generating a partial correlation coefficient where $p<0.050$.

Discussion: Although income is clearly an important resource, it cannot always be relied upon to be the permeating factor which it was expected to be. Nevertheless, income is directly linked to satisfaction with the housing environment in both the United States and Israel, though not in the Netherlands. It also emerges as an important factor in satisfaction with community services in the U.S., particularly those for which public resources fall far short of meeting demands generated by the private sector.

The fact that income exerts a strong influence on overall satisfaction with the housing environment ($p<0.010$) in both Israel and the United States stems from its compelling linkage to household responsibilities in those two nations ($p<0.005$ for both). Certainly housing costs in the San Jose SMSA are among the highest in the United States, even in the mid to late 1970s. The percentage of working women in the American sample was in excess of 55%, a fact which reflects these costs, although it by no means entirely accounts for female participation in the labor force. Similarly, the cost of housing in Israel places a severe burden on family income as costs there are comparable with those in the United States, incomes are considerably lower, and inflation far exceeds 100% per year. For example, it has been estimated that in 1975, typical Israeli families paid about five times their annual income for housing purchase, about double that usually required in the United States.[1]

In contrast, housing costs in the Netherlands have long been regulated by the public sector, particularly since the Second World War. As a result, most renters spend well under 20% of their income on housing costs, and in many cases less than 10%.[2] Consequently it is not surprising that only 27.9% of the married women in the Dutch sample were participants in the labor force. With secure tenure and regulated rents, housing satisfaction in the Netherlands has indeed been insulated from the market forces which play a far greater role in this dependent variable cluster in the United States and Israel.

In the area of community services, income again provides mixed results with respect to expectations. Only in the United States did the income factor operate as expected for the specific indexes of transportation and security ($p<0.005$ for both). These two variables are particularly sensitive to individual affluence since neither is adequately furnished by the public sector. Thus, the costs of burglar alarm systems, automatic garage door openers, and the economics of automobile ownership, operation, and maintenance clearly favor the affluent household.

However, in societies with a more robust public sector and a tradition of nationally funded local services, the distinctions created by affluence are blurred, if not submerged. Accordingly, income has no effect on satisfaction with community services in the Netherlands, and only on entertainment in Israel ($p<0.005$). Curiously, it is this variable which, along with parks, refutes expectations in the United States by exhibiting an inverse relationship to income.

The explanation for this phenomenon lies in the low-density sprawl which defines urbanization in Santa Clara County, California. Facilities and opportunities are all widely dispersed and open space, a major casualty of this development pattern, is in scarce supply. Given the region's mild climate, there exists a great potential to recreate but the demand for space and facilities far exceeds the supply. Therefore, those with the time and financial resources available for recreation and entertainment often find their needs difficult to fulfill.

With no influence on satisfaction with social patterns in California, income becomes all-pervasive in this dependent variable cluster in Israel and somewhat influential in the Netherlands. In Israel, an extremely important impact of an increase in family income is the higher incidence of automobile ownership and hence greater mobility of such families. Automobile availability greatly increases a family's ability to maintain and enrich its social lives, especially on weekends when religious laws shut down, or greatly curtail, public transportation. This may also help explain why Israeli women, when compared to their American and Dutch counterparts, registered the lowest satisfaction with their transportation facilities ($p<0.001$) as well as dissatisfaction with the quality of their entertainment ($p<0.001$) and social patterns ($p<0.050$). Lastly, income appears to have no significant effect on measures of psychological well-being in all three societies.

## 2. Education

*Expectations*: Level of education will vary positively with all dependent satisfaction variables.

*Findings*: Our data demonstrate the validity of these expectations with respect to the United States, somewhat in Israel, but not in the case of the Netherlands (see Table 4.2). All four groups of dependent variables from the American sample show some positive relationship with level of education. With respect to the housing environment, satisfaction with its appearance, an aesthetic value judgement, bears out expectations ($p<0.010$). Israel is mainly significant in the realm of psychological well-being and that from the Netherlands runs contrary, with satisfaction with privacy relating inversely to level of education ($p<0.050$).

With respect to community services, the linkage becomes more compelling in the United States but fails to lend support in Israel and in the Netherlands. The latter nation provides conflicting evidence with a negative linkage with transportation ($p<0.050$) and a positive one with satisfaction with entertainment ($p<0.050$). In Israel, only satisfaction with transportation shows even a weak correspondence with stated expectations ($p<0.100$). The American data stand in strong contrast with

TABLE 4.2 RELATIONSHIPS BETWEEN YEARS OF WOMAN'S EDUCATION AND INDEXES OF SATISFACTION: U.S.A., ISRAEL AND THE NETHERLANDS

| Satisfaction Index | Partial Correlation Coefficient[a] | | |
|---|---|---|---|
| | U.S.A. (n=825) | Israel (n=295) | Netherlands (n=215) |
| a. Housing Environment | 0.0327 | 0.0535 | -0.0867 |
| 1. Privacy | -0.0342 | 0.0357 | -0.1637 |
| 2. House & Lot | 0.0001 | 0.0699 | -0.0604 |
| 3. Household Responsibilities | -0.0110 | 0.0087 | -0.1012 |
| 4. Appearance | 0.0954[d] | 0.0341 | 0.0657 |
| b. Community Services | 0.1020[e] | -0.0555 | 0.0349 |
| 1. Parks | -0.0212 | -0.0050 | 0.0524 |
| 2. Schools | 0.0850[c] | -0.0683 | -0.0113 |
| 3. Security | 0.1442[e] | 0.0105 | 0.0465 |
| 4. Child Care | 0.0433 | -0.0062 | -0.0227 |
| 5. Transportation | 0.0303 | 0.1230[b] | -0.1833[c] |
| 6. Entertainment | 0.0580[b] | 0.0407 | 0.1661[c] |
| c. Social Patterns | 0.0705[c] | 0.0351 | -0.1227 |
| 1. Friendships | 0.0358 | 0.1254[b] | 0.0858 |
| 2. Group Activities | 0.0805[c] | 0.0312 | -0.1947[c] |
| 3. Sense of Belonging | 0.0880[c] | 0.0055 | -0.1060 |
| d. Psychological Well-being | 0.0625[b] | 0.1059 | 0.0391 |
| 1. Fullness vs. Emptiness of Life | 0.0876[c] | 0.0695 | -0.0061 |
| 2. Social Respect | 0.0797[c] | 0.3178[e] | 0.0796 |
| 3. Personal Freedom | -0.0375 | 0.0551 | 0.1574[c] |
| 4. Companionship | 0.0730[c] | 0.1517[c] | 0.0477 |
| 5. Tranquility vs. Anxiety | 0.0592[b] | -0.0178 | -0.0836 |
| 6. Self Approving vs. Guilty | -0.0021 | 0.0285 | -0.0143 |
| 7. Self Confidence | 0.0198 | 0.0686 | -0.0255 |
| 8. Elation vs. Depression | 0.0450 | 0.0478 | 0.0366 |

a. two-tailed t-test yields p>0.100 unless otherwise noted
b. t-test yields p<0.100
c. t-test yields p<0.050
d. t-test yields p<0.010
e. t-test yields p<0.005

overall satisfaction with community services providing strong confirmation with expectations ($p<0.005$) as well as satisfaction with schools ($p<0.050$) and with security ($p<0.005$). A similar set of results emerges in the social patterns dependent variable cluster. American data show a consistently strong relationship, both overall and with the specific indexes of group activities and sense of belonging (all $p<0.050$). However, in Israel no measure is sufficiently compelling to conform to expectations, and in the Netherlands, satisfaction with group activities runs contrary to expectations ($p<0.050$).

It is only in the area of psychological well-being that all three societies show some support for our stated expectations. In the United States three specific indexes (fullness vs. emptiness of life, social respect, and companionship, all relate positively to level of education, $p<0.050$) as do two in Israel (social respect, $p<0.005$, and companionship, $p<0.050$), and one in the Netherlands (personal freedom, $p<0.050$).

Discussion: While education does provide qualitative benefits to the population in the American sample, its impact is more difficult to discern among sampled populations in Israel and in the Netherlands. In order to scrutinize the impact of this variable more closely, sample populations have been split into two groups, those who have gone no further than the American equivalent of a high school diploma and those who have at least some background in higher education (see Table 4.3 for a comparison of mean scores). As can be seen, a higher level of education continues to provide higher levels of satisfaction in an impressive array of variables in the U.S. With respect to Israel the impact of education is in accordance with expectations, while in the Netherlands it does not appear to be a factor whatsoever.

In the area of psychological well-being, the mean score difference in the overall index of satisfaction is quite significant in both the U.S. and in Israel ($p<0.010$, $p<0.050$, respectively). Further, five of eight specific indexes in the American sample revealed t-tests where $p<0.050$; in Israel only one specific index (fullness vs. emptiness of life) withstood this test, while no specific indexes registered a significant statistical difference in the Netherlands. Satisfaction with social patterns can also be

TABLE 4.3 MEAN SCORES FOR INDEXES OF SATISFACTION FOR RESPONDENTS WITH HIGH SCHOOL AND COLLEGE BACKGROUNDS: U.S.A., ISRAEL AND THE NETHERLANDS

| Satisfaction Index | Satisfaction Scores[a.] | | | | | |
|---|---|---|---|---|---|---|
| | U.S.A. | | Israel | | Netherlands | |
| | HS (n=209) | Coll (n=616) | HS (n=154) | Coll (n=129) | HS (n=186) | Coll (n=29) |
| a. Housing Environment | 5.57 | 5.60 | 4.74 | 4.88 | 5.16 | 4.90 |
| 1. Privacy | 5.85 | 5.81 | 5.31 | 5.47 | 4.94 | 4.79 |
| 2. House & Lot | 5.66 | 5.53 | 4.97 | 5.23 | 5.53 | 5.17 |
| 3. Household Responsibilities | 5.08 | 5.26[b.] | 3.95 | 4.50[e.] | 4.92[b.] | 4.48 |
| 4. Appearance | 5.70 | 5.78 | 4.77[d.] | 4.32 | 5.18 | 5.10 |
| b. Community Services | 4.97 | 5.15[d.] | 4.02 | 4.18 | 5.10 | 5.29 |
| 1. Parks | 4.86 | 5.01 | 4.48 | 4.50 | 5.31 | 5.51 |
| 2. Schools | 5.02 | 5.23[b.] | 4.87 | 4.81 | 5.78 | 5.68 |
| 3. Security | 4.94 | 5.34[e.] | 4.97 | 5.34[c.] | 4.28 | 4.53 |
| 4. Child Care | 4.15 | 4.19 | 3.61 | 3.96 | 4.23 | 4.22 |
| 5. Transportation | 5.40 | 5.42 | 3.57 | 3.88 | 6.07 | 5.59 |
| 6. Entertainment | 5.23 | 5.58[e.] | 2.69 | 3.20[c.] | 4.92 | 5.59[b.] |
| c. Social Patterns | 5.18 | 5.37[c.] | 4.34 | 4.54 | 4.47 | 4.52 |
| 1. Friendships | 5.51 | 5.70[c.] | 4.70 | 4.90 | 5.21 | 5.43 |
| 2. Group Activities | 4.95 | 5.35[e.] | 3.22 | 3.88[e.] | 4.72[b.] | 4.13 |
| 3. Sense of Belonging | 5.00 | 5.05 | 5.15 | 4.97 | 4.18 | 4.00 |

| | | | | | | |
|---|---|---|---|---|---|---|
| d. Psychological Well-Being | 6.71 | 7.08[d.] | 6.78 | 7.07[c.] | 6.54 | 6.55 |
| 1. Fullness vs. Emptiness of Life | 6.93 | 7.20[c.] | 6.31 | 6.90[e.] | 6.46 | 6.24 |
| 2. Social Respect | 7.38 | 8.72 | 7.42 | 8.29[b.] | 6.78 | 6.79 |
| 3. Personal Freedom | 6.60 | 7.00d. | 6.60 | 6.95[b.] | 6.90 | 7.14 |
| 4. Companionship | 7.36 | 7.60[c.] | 7.38 | 7.59 | 7.02 | 7.10 |
| 5. Tranquility vs. Anxiety | 6.12 | 6.32[b.] | 6.39 | 6.69 | 6.10 | 6.07 |
| 6. Self Approving vs. Guilty | 6.64 | 6.79 | 7.09 | 7.02 | 6.19 | 6.17 |
| 7. Self Confidence | 6.74 | 7.28[e.] | 6.60 | 6.71 | 6.21 | 6.24 |
| 8. Elation vs. Depression | 6.51 | 6.73[c.] | 6.43 | 6.70 | 6.62 | 6.62 |

a. two-tailed t-test for all score differences yields $p>0.100$ unless otherwise noted
b. t-test for score difference yields $p<0.100$
c. t-test for score difference yields $p<0.050$
d. t-test for score difference yields $p<0.010$
e. t-test for score difference yields $p<0.005$

influenced by educational attainment, but again, the impact of this variable is greatest in the U.S. Both the overall index of satisfaction and that for satisfaction with friendships are differentiated by educational background ($p<0.050$); that for satisfaction with group activities provides even more strong support ($p<0.005$). This latter variable maintains a similar level of significance in Israel while in the Netherlands, it runs contrary to expectations ($p<0.100$). With respect to community services, level of education again remains a factor as satisfaction with entertainment and with security are significant variables in both the U.S. and in Israel ($p<0.005$ in the former, $p<0.050$ in the latter). Satisfaction with community services in the Netherlands does not conform to expectations. Only in the housing environment dependent variable cluster is education not a determinant of satisfaction, although it is linked to the financially sensitive variable of satisfaction with household responsibilities in both the United States ($p<0.100$) and in Israel ($p<0.005$). This relationship does not obtain in the Netherlands.

Thus, by this aggregation of two distinct groups of educational attainment, it can be said that this independent variable exercises an influence on the quality of life in societies where education is more avidly pursued by the populace. Nearly three-quarters (74.7%) of the American sample had at least some background in higher education, while only 13.5% of the Dutch sample did; Israel proved to be an intermediate case with somewhat less than half (43.6%) of the sample pursuing a degree beyond high school.

## 3. Occupational Status

Expectation: A positive relationship exists between occupational status and psychological well-being variables.

Findings: The tabulations of this variable do not provide strong support for expectations (see Table 4.4). Indeed, none of the specific indexes of psychological well-being demonstrated a strong statistical relationship to occupational status (where $p<0.050$) in any of the three societies studied. Also unexpected was the strong linkage of occupational status with overall satisfaction with the housing environment in the United States ($p<0.010$), a measure based on the underlying

TABLE 4.4 RELATIONSHIPS BETWEEN WOMAN'S OCCUPATION AND INDEXES OF SATISFACTION: U.S.A., ISRAEL AND THE NETHERLANDS

| Satisfaction Index | Partial Correlation Coefficient[a.] | | |
|---|---|---|---|
| | U.S.A. (n=825) | Israel (n=295) | Netherlands (n=215) |
| a. Housing Environment | 0.0926[d.] | -0.0013 | 0.0852 |
| 1. Privacy | 0.0421 | 0.0062 | 0.1450[b.] |
| 2. House & Lot | 0.0268 | -0.0265 | 0.0527 |
| 3. Household Responsibilities | 0.0713[c.] | 0.0747 | 0.1090 |
| 4. Appearance | 0.1227[e.] | -0.0625 | -0.0688 |
| b. Community Services | 0.0276 | 0.1396 | 0.0015 |
| 1. Parks | 0.0061 | 0.0135 | 0.0043 |
| 2. Schools | 0.0473 | 0.1335[b.] | 0.0716 |
| 3. Security | 0.0396 | 0.0519 | 0.0610 |
| 4. Child Care | -0.0359 | 0.0595 | 0.0414 |
| 5. Transportation | 0.0013 | -0.0455 | 0.1205 |
| 6. Entertainment | -0.0041 | 0.0830 | 0.0664 |
| c. Social Patterns | 0.0484 | 0.0242 | 0.0515 |
| 1. Friendships | 0.0337 | -0.0636 | 0.0085 |
| 2. Group Activities | 0.0632[b.] | 0.0569 | 0.1708[c.] |
| 3. Sense of Belonging | 0.0064 | -0.0006 | 0.0998 |
| d. Psychological Well-Being | -0.0139 | -0.1093 | -0.0236 |
| 1. Fullness vs. Emptiness of Life | -0.0344 | 0.0093 | 0.0131 |
| 2. Social Respect | -0.0625[b.] | -0.2830[e.] | -0.0873 |
| 3. Personal Freedom | 0.0207 | 0.0119 | -0.1657[c.] |
| 4. Companionship | -0.0336 | -0.1041 | -0.0407 |
| 5. Tranquility vs. Anxiety | 0.0260 | 0.0786 | 0.1368 |
| 6. Self Approving vs. Guilty | -0.0373 | -0.0286 | -0.0030 |
| 7. Self Confidence | 0.0013 | -0.0223 | -0.0048 |
| 8. Elation vs. Depression | 0.0093 | 0.0212 | 0.0313 |

a. two-tailed t-test yields $p>0.100$ unless otherwise noted
b. t-test yields $p<0.100$
c. t-test yields $p<0.050$
d. t-test yields $p<0.010$
e. t-test yields $p<0.005$

TABLE 4.5 MEAN SCORES OF INDEXES OF SATISFACTION FOR WOMAN'S OCCUPATIONAL STATUS: U.S.A., ISRAEL AND THE NETHERLANDS

| Satisfaction Index | Satisfaction Scores[a] | | | | | |
|---|---|---|---|---|---|---|
| | U.S.A. | | Israel | | Netherlands | |
| | non-wkg (n=405) | wkg (n=420) | non-wkg (n=94) | wkg (n=189) | non-wkg (n=155) | wkg (n=6) |
| a. Housing Environment | 5.64 | 5.54 | 4.74 | 4.83 | 5.13 | 5.10 |
| 1. Privacy | 5.85 | 5.80 | 5.35 | 5.40 | 4.96 | 4.84 |
| 2. House & Lot | 5.61 | 5.52 | 5.00 | 5.13 | 5.49 | 5.47 |
| 3. Household Responsibilities | 5.33[c] | 5.10 | 4.07 | 4.26 | 4.83 | 4.96 |
| 4. Appearance | 5.79 | 5.74 | 4.60 | 4.55 | 5.17 | 5.15 |
| b. Community Services | 5.10 | 5.10 | 4.11 | 4.08 | 5.08 | 5.22 |
| 1. Parks | 4.97 | 4.98 | 4.46 | 4.50 | 5.29 | 5.47 |
| 2. Schools | 5.19 | 5.16 | 4.64 | 4.95 | 5.67 | 6.01 |
| 3. Security | 5.26 | 5.22 | 5.03 | 5.21 | 4.24 | 4.53 |
| 4. Child Care | 4.18 | 4.20 | 3.99 | 3.68 | 4.36[b] | 3.89 |
| 5. Transportation | 5.40 | 5.42 | 3.63 | 3.74 | 5.93 | 6.19 |
| 6. Entertainment | 5.48 | 5.49 | 2.99 | 2.90 | 5.01 | 5.00 |
| c. Social Patterns | 5.35 | 5.30 | 4.19 | 4.53[c] | 4.71 | 4.70 |
| 1. Friendships | 5.64 | 5.68 | 4.48 | 4.94[d] | 5.27 | 5.18 |
| 2. Group Activities | 5.22 | 5.29 | 3.26 | 3.65[b] | 4.62 | 4.69 |
| 3. Sense of Belonging | 5.13[b] | 4.95 | 4.94 | 5.13 | 4.17 | 4.11 |

| | | | | | | |
|---|---|---|---|---|---|---|
| d. Psychological Well-Being | 6.92 | 7.07 | 6.65 | 7.04[e.] | 6.50 | 6.64 |
| 1. Fullness vs. Emptiness of Life | 7.15 | 7.12 | 6.30 | 6.72[c.] | 6.42 | 6.46 |
| 2. Social Respect | 7.44 | 7.83 | 7.30 | 8.06[c.] | 6.73 | 6.92 |
| 3. Personal Freedom | 6.76 | 7.04 | 6.57 | 6.85 | 6.92 | 6.97 |
| 4. Companionship | 7.49 | 7.58 | 7.04 | 7.70[e.] | 6.96 | 7.20 |
| 5. Tranquility vs. Anxiety | 6.28 | 6.27 | 6.17 | 6.71[d.] | 6.03 | 6.28 |
| 6. Self Approving vs. Guilty | 6.65 | 6.85[c.] | 7.06 | 7.05 | 6.17 | 6.25 |
| 7. Self Confidence | 7.02 | 7.27[d.] | 6.33 | 6.82[d.] | 6.17 | 6.33 |
| 8. Elation vs. Depression | 6.68 | 6.67 | 6.35 | 6.65[c.] | 6.57 | 6.74 |

a. two-tailed t-test for all score differences yields $p>0.100$ unless otherwise noted
b. t-test yields $p<0.100$
c. t-test yields $p<0.050$
d. t-test yields $p<0.010$
e. t-test yields $p<0.005$

positive relationships between occupational status and the specific indexes of home ownership responsibilities ($p<0.050$) and appearance ($p<0.005$). As expected, there are no significant associations between the overall indexes of satisfaction with community services or with social patterns except for satisfaction with group activities in the Netherlands ($p<0.050$).

Pursuing further the question of occupation status, mean satisfaction scores by employment status have been aggregated: women who do not work outside the home ('non-working') and those who do work outside the home ('working'). As Table 4.5 indicates, expectations are more strongly upheld, particularly in Israel, less so in the United States, and not at all in the Netherlands. In our subjective categorization, women who work outside the home are all of higher occupational status than those whose labors are confined to the domestic sphere. We find that working women are more satisfied on two specific indexes of psychological well-being in the United States, and on six indexes in Israel as well as on overall satisfaction in that realm ($p<0.005$). No satisfaction differentials were found in the Netherlands with respect to psychological well-being.

In addition to the rewards which employment has conferred in the realm of psychological well-being, Israeli working women have a more rewarding social existence than do non-working women, especially in the area of friendships ($p<0.010$) where it can be inferred that relationships may develop in a wider sphere than the geographical proximity of neighborhood contacts. Returning to Table 4.4, it may also be observed that women of higher occupational status in the Netherlands enjoy greater satisfaction with group activities ($p<0.050$). However, employment status had little or no impact on satisfaction with community services in all three nations.

Discussion: Although a statistical relationship between employment status and housing satisfaction was not anticipated, findings from Table 4.4 indicated that American women in the highest occupational echelons derived the greatest satisfaction from their housing environments. This cannot be explained by the greater financial resources conferred by this status since income was held constant in partial correlation analysis. As will be discussed later in this chapter, a more likely explanation is that American working women, especially those at the professional or managerial level, chose supportive housing environments (planned) condominiums) more often than non-working women. Thus, 47 percent of the working women in the California sample opted for multiple dwellings compared with 27 percent of the non-working women. Such a distinction could not have been made in the Netherlands or in Israel, where the entire sample lived in multiple units.

Returning to Table 4.5 it was found that working women in the United States score higher on overall housing than do their non-working counterparts. This distinction is entirely based on their higher satisfaction with household responsibilities ($p<0.050$). Since the average working woman's household income is comparable to that of the non-working woman, findings suggest she is often working to supplement the income of a spouse or is a single head of household (23% in the U.S., 3.7% in Israel, 27.9% in the Netherlands). Further, as virtually all of the American non-working women are married (95%), it seems likely that more time is available to perform household maintenance activities than for their working counterparts. The absence of the housing environment as a factor in Israel and in the Netherlands may be accounted for by the lesser demands imposed on households by multi-family living.

In summary, then, occupational status does not appear to be a major influence in this study. As with the independent variable of education, noted above, employment status' greatest impact is in the society where the greatest percentage of respondents are employed. Just as educational level was of the greatest consequence in the United States, where a much greater percentage of women pursued college training, so does occupational status appear as a more influential factor in Israel, where nearly two-thirds of the sample (66.3%) was employed, a figure which far exceeded the United States and the Netherlands (50.9% and 28%, respectively). It appears that the Israeli working woman is involved in a larger world beyond the residential neighborhood and reaps both psychological and social benefits from this activity.

## B. SUBCULTURAL INFLUENCES

### 1. Ethnic Identity

Expectation: Ethnic identity will vary inversely

with all independent variables.

Findings: This contention was only partly supported in the United States (see Table 4.6). Variables associated with the housing environment display a slight negative trend and thus weakly uphold our initial presumption. The overall index of satisfaction in this area is insignificant. However, two of the four variables in this area support our anticipation, but only one, the appearance of the housing environment ($p<0.050$), warrants statistical interest. Moving on to the community services sector, the generally weak support for expectations continues. The anticipated inverse trend seems to arise most tellingly in the score for the satisfaction with transportation variable ($p<0.050$). A similar pattern holds with social patterns variables. Of the three specific indexes within social patterns, only one supports our contention with any degree of statistical significance. The score on the group activities variable relates inversely with ethnic identity ($p<0.050$). However, with respect to psychological well-being variables, the situation changes from modest support of expectations to clear refutation. Both the overall satisfaction index and six of the eight specific indexes demonstrate strong statistical affinities which sharply cast doubt on expectations. Not only is the global score highly significant ($p<0.005$), but so is a majority of the psychological components ($p<0.050$).

Israeli and Dutch data clearly reject this expectation as there were no significant negative relationships between ethnic identity and any index of satisfaction in both countries. In Israel ethnic identity was positively related to variables concerned with the ease of household responsibilities ($p<0.050$) and overall satisfaction of community services and social patterns ($p<0.010$) especially group activities. Data from the Netherlands indicate a positive linkage between ethnic identity and indexes of child care ($p<0.050$), group activities ($p<0.050$), and receptivity towards the world ($p<0.010$).

Discussion: Although strong feelings of ethnic and/or cultural identity are not widespread (21% in our American sample) those who do identify with a particular group appear to have refuted at least some expectations. In particular, it was believed that aspects of psychological well-being would be

TABLE 4.6 RELATIONSHIPS BETWEEN ETHNIC IDENTITY AND INDEXES OF SATISFACTION: U.S.A., ISRAEL AND THE NETHERLANDS

| Satisfaction Index | Partial Correlation Coefficient[a] | | |
|---|---|---|---|
| | U.S.A. (n=825) | Israel (n=295) | Netherlands (n=215) |
| a. Housing Environment | -0.0325 | -0.0983 | -0.0541 |
| 1. Privacy | 0.0051 | 0.0884 | 0.0077 |
| 2. House & Lot | 0.0128 | -0.0055 | 0.0045 |
| 3. Household Responsibilities | -0.0145 | 0.1424[c] | -0.1293 |
| 4. Appearance | -0.0839[c] | 0.0740 | 0.0232 |
| b. Community Services | -0.0199 | 0.1443[c] | 0.1027 |
| 1. Parks | -0.0309 | 0.0170 | -0.0188 |
| 2. Schools | -0.0044 | -0.0348 | 0.0076 |
| 3. Security | 0.0431 | -0.0126 | -0.0270 |
| 4. Child Care | -0.0158 | 0.1046 | 0.1633[c] |
| 5. Transportation | -0.0716[c] | -0.0018 | -0.0220 |
| 6. Entertainment | -0.0527 | 0.1059 | 0.1079 |
| c. Social Patterns | -0.0524 | 0.1457[c] | 0.0598 |
| 1. Friendships | 0.0265 | 0.0810 | -0.0515 |
| 2. Group Activities | -0.0705[c] | 0.1287[b] | 0.1712[c] |
| 3. Sense of Belonging | -0.0063 | 0.0328 | 0.0189 |
| d. Psychological Well-Being | 0.1228[e] | -0.1104 | 0.0365 |
| 1. Fullness vs. Emptiness of Life | 0.0361 | 0.0985 | 0.1406[b] |
| 2. Social Respect | 0.1036[e] | 0.0133 | -0.0235 |
| 3. Personal Freedom | 0.0743[c] | 0.0756 | -0.0845 |
| 4. Companionship | 0.1137[e] | -0.0656 | -0.0169 |
| 5. Tranquility vs. Anxiety | 0.0837[c] | 0.0149 | 0.0849 |
| 6. Self Approving vs. Guilty | 0.1015[e] | -0.0953 | 0.0937 |
| 7. Self Confidence | 0.0902[c] | 0.0051 | -0.0286 |
| 8. Elation vs. Depression | 0.0300 | -0.0359 | 0.0358 |

a. two-tailed t-test yields p>0.100 unless otherwise noted
b. t-test yields p<0.100
c. t-test yields p<0.050
d. t-test yields p<0.010
e. t-test yields p<0.005

adversely affected by a strong feeling of ethnic identity. To the contrary, the scores indicated that ethnic identity is quite positively linked with both the general and many specific indexes in this area. As Glazer and Moynihan's classic study of New York City ethnicity indicates, identification with one's heritage in a world 'beyond the melting pot' appears to be a positive force in the lives of survey respondents.[3] Obviously, strong feelings of ethnic and/or cultural identity are a factor to be reckoned with in a socially heterogeneous society and call attention to the subtle and hidden imperatives of these feelings. As Fischer suggests in his study of California personal networks, with increases in urbanism, ethnic involvement may be more profound.[4]

However, in more tangible areas of inquiry expectations are realized. With respect to the housing environment, the specific index of the house and neighborhood appearance confirmed our anticipation of an inverse relationship ($p<0.050$) (see Table 4.6). The aesthetic homogeneity of many suburban neighborhoods may be quite different from the intensity of the housing environments that some respondents may have experienced at an earlier time in their lives. The specific index of transportation was also a source of evidence for an inverse relationship between ethnicity and all dependent variables. It could be argued that the generally low-density setting which characterizes the San Jose metropolitan area serves as an impediment to an efficient transportation system, a situation which contrasts with that found in older urban centers, particularly on the east coast, where good public transportation has long been provided.

As in the American sample, only a minority of respondents in Israel and the Netherlands (14 and 20 percent respectively) expressed a strong ethnic or cultural identity. Unlike the mixed pattern experienced by their American counterparts, individuals with strong ethnic identities in Israel and the Netherlands had only positive interactions with various aspects of their lives. Israeli women with strong ethnic identities experienced a great ease of household responsibiliites and rewarding social patterns which seem to reflect the vigorous support network provided by a number of immigrant ethnic groups which make up Israeli society. This may be especially true for Jewish families from Asia and North Africa (Sephardim) who settled in Israel relatively recently. These are characterized by extensive family networks and by traditional home-centered roles for women. These same respondents, many of whom grew up in older, deteriorated neighborhoods with poor public facilities, seemed to be satisfied with community services in their middle class housing environment.

In the Netherlands, social support networks were observed which helped create positive experiences for women in the areas of child care, group activities, and fullness of life. Here the support seemed to be generated primarily from formal institutions such as church groups and to some extent family ties. Unlike their American and Israeli counterparts, Dutch women have experienced relative geographic residential stability which tends to promote the continuity of friendships and group ties on an intergenerational basis.

In order to explore the concept of ethnic identity further, Table 4.7 offers a comparison between those with weak or no ethnic identity and those who registered strong feelings. As expected, it appears that the American respondents with a strong ethnic identity experience fewer interactions close to home. This indeed may be their preference. They have a lesser number of acquaintances and neighborly contacts within the proximate residential environment. When five neighborhood-based variables which illustrate this situation are considered, three show decisive differences ($p<0.001$) which place strong ethnics at somewhat of a disadvantage. However, when non-spatial activities are viewed, (e.g., religious and political organizations, and groups outside the neighborhoods) strong ethnics show a much greater degree of involvement. If they are less satisfied with local activities, it is evidently because of the distance factor in their non-spatial character. This accords with what is observed about California lifestyles and urban development. That is, spatial factors may be assuming less relevance in post-1945 urbanization than in earlier periods.

In contrast, Israeli and Dutch respondents with strong ethnic identities did not experience less social interaction close to home. In the case of Israeli women, strong ethnicity was actually an advantage for two of the four scored variables concerned with nearby friendships ($p<0.100$), as well as with informal group activity outside the neighborhood ($p<0.001$), and was certainly an advantage for psychological well-being ($p<0.050$). While the Dutch women having strong ethnic identities did not report the neighborhood social

TABLE 4.7 MEAN SATISFACTION SCORES FOR SELECTED VARIABLES BY RESPONDENT'S STRENGTH OF ETHNIC IDENTITY[f]: U.S.A., ISRAEL AND THE NETHERLANDS

| Variable | Satisfaction Scores[a] | | | | | |
|---|---|---|---|---|---|---|
| | U.S.A. | | Israel | | Netherlands | |
| | Weak Identity (n=717) | Strong Identity (n=96) | Weak Identity (n=240) | Strong Identity (n=55) | Weak Identity (n=172) | Strong Identity (n=43) |
| Length of Residency (years) | 3.22 | 3.64 | 3.88 | 6.25[c] | 7.26 | 7.70 |
| Speaking acquaintances (in neighborhood) | 24.39[e] | 15.45 | - | - | - | - |
| Neighbors to discuss general interest topics | 9.34[c] | 6.88 | 4.33 | 5.10 | 3.17 | 3.33 |
| Neighbors to borrow from | 4.44[e] | 3.28 | 3.08 | 3.83[b] | 3.19 | 3.79 |
| Neighbors to aid in crisis | 4.89[e] | 3.27 | 2.92 | 4.23[b] | 3.27 | 3.86 |
| Neighbors to aid in projects | 2.60[b] | 2.06 | 1.95 | 2.48 | 1.41 | 1.84 |
| Close friends one hour away | 4.94 | 7.32[b] | 4.48 | 3.98 | 1.69 | 2.76[b] |
| Strength of involvement with: | | | | | | |
| Religious groups | 2.58 | 3.11[c] | 1.12 | 1.49 | 1.33 | 3.56[e] |
| Political organizations | 1.63 | 1.97[b] | 1.27 | 1.45 | 1.24 | 1.44 |
| Informal groups outside neighborhood | 2.70 | 3.53[e] | 2.10 | 3.41[e] | 1.39 | 1.30 |
| Action groups outside neighborhood | 1.73 | 2.63[e] | 1.24[d] | 1.05 | - | - |
| Social organizations outside neighborhood | 1.86 | 2.72[e] | - | - | - | - |
| Other organized groups | 1.95 | 2.52[e] | 1.28 | 1.62 | 1.48 | 2.42[e] |
| Psychological Well-being | 6.71 | 7.09[e] | 6.53 | 6.98[c] | 6.52 | 6.59 |

a. two-tailed t-test for all score differences yields $p>0.100$ unless otherwise noted
b. t-test for score differences yields $p<0.100$
c. t-test for score differences yields $p<0.050$
d. t-test for score differences yields $p<0.010$
e. t-test for score differences yields $p<0.005$
f. Based on the question, 'Do you identify with any ethnic or cultural group?' (See questionnaire Appendix A). A 'non' or 'weak' identity is a score of 1-4; a 'strong identity is a score of 5-7

disadvantages experienced by their American counterparts, they did experience a substantial positive social interaction outside the neighborhood somewhat like American women with respect to friendships ($p<0.100$) and formal group activities, especially those of a religious nature ($p<0.001$).

In sum, the dispersed ethnicity in post-1945 suburban communities in California appears to contradict the work of Glazer and Moynihan in New York with that city's contrasting and more spatially-oriented ethnic demarcations. At the same time there appears to be a spatial social compensation at the neighborhood level for the vast majority of Americans without strong ethnic group identity. For them there is a need for a strong local support system in order to counterbalance social interactions enjoyed outside the neighborhoods by their more ethnic colleagues. In smaller nations, such as Israel and the Netherlands, such local compensation does not seem to be needed for non-ethnics as much, due to the shorter distances between successive residences and the typically fewer moves made by households. Since an overwhelming majority of the Israeli and Dutch women interviewed lived within an hour's drive from their childhood neighborhoods, a viable long-term social support network seems to be readily available to most residents, lessening the need for high levels of local socialization for non-ethnically oriented women.

## 2. Regional Origin

Expectation: All satisfaction variables will be positively related to the degree of urbanness (i.e. population size) of the community in which respondent's formative years were spent; and negatively related to the distance of that community from her current residence.

Findings: With respect to the degree of urbanness - the population size of the community of origin - our results provide some support for expectations in the United States, somewhat less in Israel and almost no support in the Netherlands. Tables 4.8, 9 and 10 illustrate these findings and differ from other data presentations in three ways. First, note that the vertical columns juxtapose communities of origin by their degree of urbanness, but only those which provide statisticaly useful information (when $p<0.100$). Thus, suburban-urban comparisons are

TABLE 4.8 MEAN SCORES OF INDEXES OF SATISFACTION FOR DEGREE OF URBANNESS[f].: U.S.A.

| SATISFACTION INDEX | RURAL- (n=78) | SMALL TOWN (n=286) | RURAL- (n=78) | SUBURBAN (n=268) | RURAL- (n=78) | URBAN (n=191) | SMALL TOWN (n=286) | SUBURBAN (n=268) | SMALL TOWN (n=286) | URBAN (n=191) |
|---|---|---|---|---|---|---|---|---|---|---|
| Housing Environment | | a. | | a. | | a. | | a. | | a. |
| Privacy | | a. | | a. | 5.69 | 5.91b. | | a. | | a. |
| House and lot | 5.34 | 5.67c. | | a. | | a. | 5.67b. | 5.49 | | a. |
| Household Responsibilities | | a. | | a. | | a. | | a. | | a. |
| Appearance | | a. | | a. | | a. | | a. | | a. |
| Community Services | 4.91 | 5.13c. | 4.91 | 5.15c. | | a. | | a. | | a. |
| Parks | | a. | | a. | | a. | | a. | | a. |
| Schools | | a. | | a. | | a. | | a. | | a. |
| Security | 4.87 | 5.32d. | 4.87 | 5.33d. | 4.87 | 5.19b. | | a. | | a. |
| Child Care | | a. | | a. | | a. | | a. | | a. |
| Transportation | | a. | | a. | | a. | | a. | | a. |
| Entertainment and culture | | a. | | a. | | a. | | a. | 5.59b. | 5.34 |
| Social Patterns | | a. | | a. | | a. | | a. | | a. |
| Friendships | 5.39 | 5.71c. | 5.39 | 5.66b. | 5.39 | 5.69b. | | a. | | a. |
| Group Activities | | a. | | a. | | a. | | a. | | a. |
| Sense of Belonging | | a. | | a. | | a. | 5.17c. | 4.90 | | a. |

a. two-tailed t-test for all score differences yields $p>0.100$ unless otherwise noted
b. t-test yields $p<0.100$
c. t-test yields $p<0.050$
d. t-test yields $p<0.010$
e. t-test yields $p<0.005$
f. Based on the question, 'Do you identify with any ethnic or cultural group?' (See questionnaire Appendix A). A 'non' or 'weak' identity is a score of 1-4; a 'strong identity is a score of 5-7

TABLE 4.9 MEAN SCORES OF INDEXES OF SATISFACTION FOR DEGREE OF URBANNESS[f]: ISRAEL

| SATISFACTION INDEX | RURAL- (n=46) | SMALL TOWN (n=64) | RURAL- (n=46) | SUBURBAN (n=40) | RURAL- (n=46) | URBAN (n=145) | SMALL TOWN (n=64) | SUBURBAN (n=40) | SMALL TOWN (n=64) | URBAN (n=145) |
|---|---|---|---|---|---|---|---|---|---|---|
| Housing Environment | a. | | a. | | a. | | a. | | a. | |
| Privacy | a. | | a. | | a. | | a. | | a. | |
| House and lot | a. | | a. | | a. | | 5.28[b] | 4.73 | a. | |
| Household Responsibilities | a. | | a. | | a. | | a. | | a. | |
| Appearance | a. | | a. | | a. | | a. | | a. | |
| Community Services | a. | | a. | | a. | | a. | | a. | |
| Parks | a. | | 4.31 | 5.04[b] | a. | | 4.24 | 5.04[c] | a. | |
| Schools | a. | | a. | | a. | | 5.20[b] | 4.53 | a. | |
| Security | a. | | a. | | a. | | a. | | 4.86 | 5.26[b] |
| Child Care | a. | | a. | | a. | | a. | | 4.10[b] | 3.56 |
| Transportation | a. | | a. | | a. | | a. | | 4.04[b] | 3.51 |
| Entertainment and culture | a. | | a. | | a. | | a. | | a. | |
| Social Patterns | a. | | a. | | a. | | a. | | a. | |
| Friendships | 4.54 | 4.98[b] | a. | | a. | | a. | | a. | |
| Group Activities | a. | | a. | | a. | | a. | | a. | |
| Sense of Belonging | a. | | a. | | a. | | a. | | a. | |

a. two-tailed t-test for all score differences yields $p>0.100$ unless otherwise noted
b. t-test yields $p<0.100$
c. t-test yields $p<0.050$
d. t-test yields $p<0.010$
e. t-test yields $p<0.005$
f. Based on the question, 'Do you identify with any ethnic or cultural group?' (See questionnaire Appendix A). A 'non' or 'weak' identity is a score of 1-4; a 'strong'identity is a score of 5-7

TABLE 4.10 MEAN SCORES OF INDEXES OF SATISFACTION FOR DEGREE OF URBANNESS[f]: THE NETHERLANDS

| SATISFACTION INDEX | RURAL- (n=21) | SMALL TOWN (n=19) | RURAL- (n=21) | SUBURBAN (n=18) | RURAL- (n=21) | URBAN (n=157) | SMALL TOWN (n=19) | SUBURBAN (n=18) | SMALL TOWN (n=19) | URBAN (n=157) |
|---|---|---|---|---|---|---|---|---|---|---|
| Housing Environment | a. | | a. | | a. | | a. | | a. | |
| Privacy | a. | | a. | | a. | | a. | | a. | |
| House and lot | a. | | a. | | a. | | a. | | a. | |
| Household Responsibilities | a. | | 5.08[b] | 4.53 | a. | | a. | | a. | |
| Appearance | a. | | a. | | 4.50 | 5.26[c] | a. | | a. | |
| Community Services | a. | | a. | | a. | | a. | | a. | |
| Parks | a. | | a. | | a. | | a. | | a. | |
| Schools | a. | | a. | | a. | | a. | | a. | |
| Security | a. | | a. | | a. | | a. | | a. | |
| Child Care | a. | | a. | | a. | | a. | | a. | |
| Transportation | 6.29[c] | 5.69 | a. | | a. | | a. | | a. | |
| Entertainment and culture | a. | | a. | | a. | | a. | | a. | |
| Social Patterns | a. | | a. | | a. | | a. | | a. | |
| Friendships | a. | | a. | | a. | | a. | | a. | |
| Group Activities | a. | | a. | | a. | | a. | | 5.11[b] | 4.56 |
| Sense of Belonging | a. | | a. | | a. | | a. | | a. | |

a. two-tailed t-test for all score differences yields $p>0.100$ unless otherwise noted
b. t-test yields $p<0.100$
c. t-test yields $p<0.050$
d. t-test yields $p<0.010$
e. t-test yields $p<0.005$
f. Based on the question, 'Do you identify with any ethnic or cultural group?' (See questionnaire Appendix A). A 'non' or 'weak' identity is a score of 1-4; a 'strong' identity is a score of 5-7

omitted. Second, because no significant relationships exist between the overall index of satisfaction and specific indexes of satisfaction in the area of psychological well-being, those variables have been deleted. Third, the actual presentation of the data itself will only include those mean scores that lend themselves to a statistically relevant analysis. Thus, any scores which result in $p>0.100$ will be designated by the superscript 'a.'.

In Table 4.8, it is clear that American respondents from the least urban of origins, those from rural areas, exhibit the lowest levels of satisfaction with the housing environment; two of its four specific indexes are of statistical interest. Respondents with urban origins were somewhat more satisfied with the specific index of privacy ($p<0.100$) than those of rural background. Rural respondents also fare less well than those of small town background in the area of house and lot satisfaction ($p<0.050$). The pattern of statistically significant scores demonstrates that, when compared with those of small town and suburban origins, respondents of rural background tend to be least satisfied, continues in the area of community services, both in the overall index of satisfaction and that with the specific index of security ($p<0.010$). Interestingly, even respondents from urban areas score higher in this latter area ($p<0.100$). The pattern also holds when social patterns variables are considered. The specific index of satisfaction with friendships shows those of rural background to be the least satisfied when juxtaposed with the other three levels of urbanness. Respondents from small towns fared better ($p<0.050$) as did those from suburban and urban backgrounds ($p<0.100$ for both). Thus, it is in conformance with expectations that respondents from rural backgrounds will tend to be the least satisfied with their current suburban situations.

But what of juxtapositions between respondents of small town backgrounds versus those from suburban and urban backgrounds? Here the evidence departs from expectations. There are two specific indexes (house and lot, and sense of belonging) which show statistically relevant results for small town-suburban comparisons. In both, scores for women from small town backgrounds are higher ($p<0.100$, $p<0.050$, respectively). Similarly, on the one specific index where a statistically significant relationship exists between small town and urban-originated populations, satisfaction with entertainment and culture, respondents from a small town background fared a bit better ($p<0.100$). With respect to comparisons between women from suburban and urban backgrounds there were, as previously stated, no statistically relevant comparisons to be made.

Our Israeli data, shown on Table 4.9, reveals a similar but much less pronounced pattern of positive relationships between the degree of urbanness and quality of life satisfaction. We find only a muted disadvantage of rural backgrounds with regard to the satisfaction of park use and friendships (both $p<0.100$). With regard to small town backgrounds, results are mixed with some small town advantage. When compared to their counterparts with suburban origins, Israeli women from small towns registered satisfactions which were higher for house and lot, and schools (both $p<0.100$) and lower for parks ($p<0.050$). When matched with women from urban areas, respondents from rural backgrounds scored modest advantages in the areas of child care and transportation (both $p<0.100$) and a slight disadvantage about security ($p<0.100$). There were no significant relationships between small town backgrounds and social pattern variables.

Results from the Netherlands provided data which mildly rejected expectations about urbanness. As Table 4.10 reveals, women with rural backgrounds expressed significant satisfaction with household responsibilities ($p<0.100$) and transportation ($p<0.050$), while registering a dissatisfaction with housing appearance ($p<0.050$). Although not as pronounced as that of their American and Israeli counterparts, Dutch respondents with small town origins had a modest advantage in the area of group activities ($p<0.100$).

The question pertaining to the distance of one's location of origin has yielded results that provide weak support for expectations in the United States, mixed support in Israel, and no impact in the Netherlands (see Table 4.11). In the United States variables pertaining to the housing environment support expectations. Not only is the global score in line with what was anticipated ($p<0.050$), but so is the appearance of the housing environment ($p<0.100$). Running counter to this, however, is satisfaction with house and lot which shows an unexpected positive relationship ($p<0.050$). With respect to community services, there is a general show of support, but only the satisfaction

TABLE 4.11 RELATIONSHIPS BETWEEN LOCATION OF ORIGIN AND INDEXES OF SATISFACTION: U.S.A., ISRAEL AND THE NETHERLANDS

| Satisfaction Index | Partial Correlation Coefficient[a] | | |
|---|---|---|---|
| | U.S.A. (n=825) | Israel (n=295) | Netherlands (n=215) |
| a. Housing Environment | -0.0780[c] | -0.0223 | -0.1004 |
| 1. Privacy | -0.0347 | -0.0638 | -0.0650 |
| 2. House & Lot | 0.0811[c] | -0.0709 | -0.0584 |
| 3. Household Responsibilities | -0.0329 | 0.0889 | -0.0493 |
| 4. Appearance | -0.0594[b] | -0.0129 | -0.0908 |
| b. Community Services | -0.0333 | -0.1394[b] | -0.0424 |
| 1. Parks | -0.0300 | -0.0710 | -0.0515 |
| 2. Schools | -0.0194 | -0.0369 | 0.0727 |
| 3. Security | 0.0104 | -0.0817 | -0.0323 |
| 4. Child Care | -0.0730[c] | 0.1269[b] | 0.0243 |
| 5. Transportation | -0.0222 | 0.0292 | 0.0313 |
| 6. Entertainment | -0.0035 | 0.0600 | -0.0134 |
| c. Social Patterns | -0.0406 | 0.0120 | -0.0159 |
| 1. Friendships | -0.0612[b] | -0.0672 | -0.0340 |
| 2. Group Activities | -0.0178 | 0.1170[b] | -0.0495 |
| 3. Sense of Belonging | -0.0240 | -0.0127 | 0.0209 |
| d. Psychological Well-Being | 0.0459 | -0.0197 | -0.1133 |
| 1. Fullness vs. Emptiness of Life | -0.0081 | 0.1180[b] | -0.1050 |
| 2. Social Respect | 0.0952[d] | 0.3466[e] | -0.0510 |
| 3. Personal Freedom | 0.0270 | 0.0268 | -0.1140 |
| 4. Companionship | 0.0184 | -0.0323 | -0.0311 |
| 5. Tranquility vs. Anxiety | 0.0447 | -0.0975 | -0.0964 |
| 6. Self Approving vs. Guilty | -0.0184 | 0.0085 | -0.0720 |
| 7. Self Confidence | 0.0170 | 0.0907 | -0.0431 |
| 8. Elation vs. Depression | -0.0305 | -0.0071 | -0.0534 |

a. two-tailed t-test yields $p>0.100$ unless otherwise noted
b. t-test yields $p<0.100$
c. t-test yields $p<0.050$
d. t-test yields $p<0.010$
e. t-test yields $p<0.005$

with child care variable shows statistical significance ($p<0.050$). The group of variables under the general heading social patterns continues this trend. Although only one variable, friendships, shows marginal statistical significance ($p<0.100$), the consistent inverse relationship between these dependent variables and locational origins lends a degree of credence to expectations. Less successful in fulfilling presumptions is the cluster of psychological well-being specific indexes, which fail to show a compelling statistical relationship with the distance of respondents' locations of origin.

Israeli results are less decisive than those of the United States. Israeli support for expectations was shown with the weak negative relationships between distance of one's location of origin and overall satisfaction of community services, group activities, and fullness of life (all $p<0.100$). At the same time, Israeli rejection of this contention is shown with the positive relationships between location of origin and satisfaction with child care ($p<0.100$) and social respect ($p<0.001$).

While in the Netherlands there was generally a directional support for expectations, none of the relationships between the distance of one's location of origin and quality of life factors was statistically significant.

Discussion: It is clear that American respondents from rural areas are notably less satisfied with many aspects of their lives than are those from more urbanized origins. Measures of satisfaction with security and with friendships are the most telling. Concern with rising rates of criminal behavior is pandemic in many metropolitan areas, especially as the social and economic characteristics of suburban populations continue to broaden. Perhaps rurally-raised respondents are experiencing more difficulty in adapting to both the positive and negative aspects of the urbanization of the suburbs. They are also less satisfied with the friendships which they have experienced. If, as Fischer discovered, less-urban people are more dependent on kinship ties for friendships than are city residents, a move of such individuals away from family to a large metropolitan area may create a more difficult social adjustment for them.[5] Thus, respondents from rural backgrounds tend to feel less materially secure, but also perhaps less socially secure as well. On the other hand, there is no solid evidence that the increasing urbanness of one's origins produces greater levels of satisfaction once the rural population is excluded. As indicated, there are no statistically significant relationships between respondents of suburban and urban backgrounds. With limited opportunities to evaluate small town versus suburban and urban backgrounds, the evidence - albeit limited - suggests that small town backgrounds tend to produce greater levels of satisfaction. Is it a matter of expectations which are formed during one's formative years? The limitations imposed by small town life are well-known - although Americans today rate the qualities of such an environment rather highly as many opinion polls have shown.[6] Perhaps what renders this trend less surprising is the fact that the U.S. sample population contains a plurality from a small town background, 34.8%; with 32.6% from suburbs, 23.2% from large cities and 9.5% from rural areas. Perhaps suburban life is the ideal compromise for those seeking small town ambience while at the same time having access to metropolitan economic opportunities.

The disadvantages of a rural background for women living in a metropolitan setting were far less clear in Israel and in the Netherlands than in the United States. Israeli respondents with rural origins expressed a weak dissatisfaction with parks and friendship patterns while their Dutch counterparts actually registered a slight advantage of rural upbringing for transportation satisfaction as well as mixed results about the quality of housing. This blurring of differences of the influence of the degree of urbanness is also shown by the relatively mixed or weak impact of small town background on satisfaction scores in Israel and Holland. While lifestyles could vary substantially between urban and rural areas in these countries, especially in Israel with its array of collective settlements, the relatively small size of these nations brings an overwhelming majority of their rural populations within one to two hours from a major metropolitan area. Thus, it could be argued that most individuals in metropolitan areas with rural backgrounds in Israel and the Netherlands require less of an adjustment than their American counterparts since they have already lived within the orbit of these urbanized areas and can maintain many linkages to their home regions.

The weak support found for the influence of the distance of one's origins on the quality of metropolitan life may be partially due to the peculiarities and technology of long distance travel. For example, the sheer mileage as the crow flies may not alone be the best source of a productive inquiry. Perhaps more important is the ease and/or travel time involved in making a return trip to the source of one's roots. It may be much easier to fly from San Jose to a major eastern metropolitan area than it would be to reach a smaller community one-third to one-half the distance. An airline connection to Wichita, Des Moines, or Amarillo may be a far greater ordeal than transportation to the Atlantic coast. In another instance in Israel, distance from one's origins actually had a strong positive impact on social respect, probably because respondents furthest removed from their home origins are usually from western countries and are better received immigrants in the European (Ashkenazy) dominated Israeli society than are immigrants from nearby middle-eastern nations.

As was the case for degree of urbanism, the limited impact of the location of origin on satisfaction scores may have much to do with what appears to be a social convergence of regional lifestyles due to increased communication and transportation facilities, even in a geographically large country like the United States.[7] For example, John Palen believes that urban and rural definitions lose their explanatory value in a time of national metropolitan societies.[8] Thus, if regions are in fact becoming more similar, the impact of growing up in one area and moving to another should generally decline.

3. Length of Residency

Expectation: The length of residency will be positively related to all dependent satisfaction variables.

Findings: Results provided modest support of expectations in the United States and very limited support in the Netherlands and in Israel. As Table 4.12 reveals, with respect to the housing environment, results do not support this expectation in the United States. The overall trend is negative with one specific index, appearance of housing environment, showing an inverse relationship ($p<0.050$). The community services cluster of variables reverses this trend somewhat, but only

TABLE 4.12 RELATIONSHIPS BETWEEN LENGTH OF LOCAL RESIDENCY AND INDEXES OF SATISFACTION: U.S.A., ISRAEL AND THE NETHERLANDS

| Satisfaction Index | Partial Correlation Coefficient[a] | | |
|---|---|---|---|
| | U.S.A. (n=825) | Israel (n=295) | Netherlands (n=215) |
| a. Housing Environment | -0.0462 | -0.0637 | 0.0601 |
| 1. Privacy | -0.0548 | -0.0108 | 0.0526 |
| 2. House & Lot | -0.0026 | -0.0621 | -0.0946 |
| 3. Household Responsibilities | -0.0226 | -0.0087 | 0.2190[d] |
| 4. Appearance | -0.0626[c] | -0.0380 | -0.0000 |
| b. Community Services | 0.0051 | -0.0774 | 0.1360 |
| 1. Parks | -0.0218 | -0.1087 | 0.0238 |
| 2. Schools | 0.0093 | -0.0422 | 0.0603 |
| 3. Security | -0.0543 | -0.0341 | -0.0936 |
| 4. Child Care | 0.0260 | -0.0369 | 0.1318 |
| 5. Transportation | 0.1183[e] | 0.0109 | 0.0167 |
| 6. Entertainment | 0.0393 | 0.1166[b] | 0.1184 |
| c. Social Patterns | 0.0722[c] | 0.0695 | 0.0822 |
| 1. Friendships | 0.0718 | 0.0981 | 0.0830 |
| 2. Group Activities | -0.0280 | -0.0058 | 0.0649 |
| 3. Sense of Belonging | 0.1251[e] | 0.0624 | 0.0546 |
| d. Psychological Well-Being | 0.0008 | -0.0540 | 0.0170 |
| 1. Fullness vs. Emptiness of Life | 0.0529 | 0.0088 | 0.0291 |
| 2. Social Respect | -0.0241 | -0.0644 | -0.0180 |
| 3. Personal Freedom | 0.0029 | -0.1368[c] | 0.0885 |
| 4. Companionship | -0.0512 | 0.0486 | 0.0366 |
| 5. Tranquility vs. Anxiety | 0.0234 | -0.0044 | -0.0516 |
| 6. Self Approving vs. Guilty | 0.0146 | 0.0276 | 0.0421 |
| 7. Self Confidence | 0.0124 | -0.0594 | 0.0280 |
| 8. Elation vs. Depression | 0.0369 | -0.0444 | -0.0687 |

a. two-tailed t-test yields $p>0.100$ unless otherwise noted
b. t-test yields $p<0.100$
c. t-test yields $p<0.050$
d. t-test yields $p<0.010$
e. t-test yields $p<0.005$

one, satisfaction with transportation, conforms to what was anticipated ($p<0.005$). The area which shows the strongest support for expectations is that of social patterns. Both the overall index of satisfaction and the specific index of friendships are statistically significant ($p<0.050$) while the sense of belonging variable is a further source of support ($p<0.005$). The psychological well-being cluster of variables as a group shows virtually no relationship with length of residency. Thus, social patterns is the only cluster of variables which seems to support expectations in the United States.

Israeli and Dutch data indicate that the influence of length of residency had an even less pronounced impact on quality of life factors than that in the United States. Israeli women reported a mildly positive benefit of length of local residency with respect to entertainment and culture ($p<0.100$) and a negative association regarding their personal freedom ($p<0.050$). In the Netherlands length of residency had only one significant relationship, a positive linkage with satisfaction of household responsibilities ($p<0.050$).

If this examination of Israel and the Netherlands is broadened to include length of national residency, there exists mixed support for this contention. As Table 4.13 shows, Israeli women indicated that with an increase of national residency, they were less satisfied with security ($p<0.050$) and fullness of life ($p<0.100$), and much more satisfied with the degree of social respect ($p<0.001$). The Dutch women with the longest national residency experienced a decline in housing privacy ($p<0.050$) and, as in Israel, had less satisfaction with security ($p<0.001$).

Discussion: In industrialized nations, does familiarity breed contempt? Or, are expectations inevitably subject to downward revision or disappointment? These questions appear to be mildly operative in the case of the housing environment in the United States locally, and the Netherlands nationally. The inverse relationship between the index of satisfaction of privacy and the appearance of the housing environment and length of residency may be a signal in these countries. There was also an indication of a definite decline in satisfaction of neighborhood security with an increase in national residency in Israel and in the Netherlands. Perhaps the major source of dissatisfaction with the passage of time is with the symbolic and security

TABLE 4.13 RELATIONSHIPS BETWEEN LENGTH OF NATIONAL RESIDENCY AND INDEXES OF SATISFACTION: ISRAEL AND THE NETHERLANDS

| Satisfaction Index | Partial Correlation Coefficient[a] | |
|---|---|---|
| | Israel (n=295) | Netherlands (n=215) |
| a. Housing Environment | -0.0419 | -0.0796 |
| 1. Privacy | -0.0391 | -0.1692[c] |
| 2. House & Lot | -0.0458 | -0.0113 |
| 3. Household Responsibilities | 0.0001 | -0.0065 |
| 4. Appearance | -0.0161 | -0.0129 |
| b. Community Services | -0.1508[c] | -0.0684 |
| 1. Parks | -0.0356 | -0.0187 |
| 2. Schools | 0.0692 | 0.0334 |
| 3. Security | -0.1386[c] | -0.2219[e] |
| 4. Child Care | 0.0214 | 0.0569 |
| 5. Transportation | 0.0208 | 0.0891 |
| 6. Entertainment | 0.0492 | -0.0930 |
| c. Social Patterns | -0.0944 | -0.0324 |
| 1. Friendships | -0.0253 | -0.0313 |
| 2. Group Activities | 0.0315 | -0.0167 |
| 3. Sense of Belonging | -0.1040 | 0.0120 |
| d. Psychological Well-Being | -0.0549 | 0.0300 |
| 1. Fullness vs. Emptiness of Life | -0.1272[b] | -0.0524 |
| 2. Social Respect | 0.2238[e] | 0.1460[b] |
| 3. Personal Freedom | 0.0900 | 0.0307 |
| 4. Companionship | -0.0177 | 0.0094 |
| 5. Tranquility vs. Anxiety | -0.0712 | -0.0016 |
| 6. Self Approving vs. Guilty | -0.0795 | -0.0081 |
| 7. Self Confidence | -0.0149 | 0.0165 |
| 8. Elation vs. Depression | -0.0560 | 0.0302 |

a. two-tailed t-test yields $p>0.100$ unless otherwise noted
b. t-test yields $p<0.100$
c. t-test yields $p<0.050$
d. t-test yields $p<0.010$
e. t-test yields $p<0.005$

properties of the house and neighborhood. Although the age of the dwelling unit will inevitably result in increased maintenance responsibilities, the negative scores on this item were not significant, and in the Dutch case was positive, a factor which reflects lower costs for older housing. Perhaps, then, the discontent which accumulates over time is not strongly focussed, and, as in the Netherlands, may be countervailed by attractive pricing for housing.

On the other hand, social patterns do show strong support for one supposition in the United States. This is clearly a matter of common sense, for no matter how satisfactory or unsatisfactory a living environment, close friends will be made with the passage of time. One's sense of belonging makes the strongest statement of this phenomenon. Yet, it is curious that there are no strong American correlations in the psychological realm.

In Israel and the Netherlands the only psychological benefit of length of residency was social respect, which suggests the granting of status to the more senior members of a community, not uncommon in tradition-oriented societies. At the same time, Israeli women expressed a psychological dissatisfaction in terms of loss of personal freedom and fullness of life with increased residency, a condition which reflects the great pressures of living in this politically, economically and militarily beleaguered nation. The negative scores for security may also reflect these pressures as well as a general decline in a feeling of safety in metropolitan areas. Indeed, Dutch women also expressed a negative association between security and length of residency, and registered the lowest score on security compared to the other two countries studied ($p<0.001$). Perhaps the social respect accorded long-time residents represents a social reward for dealing with the changing metropolitan areas and for holding the community fabric together.

## C. LIFE CYCLE INFLUENCES

### 1. Age

Expectation: Age will vary positively with satisfaction with the housing environment and negatively with psychological well-being.

Findings: As indicated in Table 4.14, survey data

TABLE 4.14 RELATIONSHIPS BETWEEN WOMAN'S AGE AND INDEXES OF SATISFACTION: U.S.A., ISRAEL AND THE NETHERLANDS

| Satisfaction Index | Partial Correlation Coefficient[a.] | | |
|---|---|---|---|
| | U.S.A. (n=825) | Israel (n=295) | Netherlands (n=215) |
| a. Housing Environment | 0.1063[e.] | 0.0291 | 0.0211 |
| 1. Privacy | 0.0421 | 0.0245 | 0.0420 |
| 2. House & Lot | 0.0659[b.] | 0.1043 | 0.0393 |
| 3. Household Responsibilities | 0.0355 | -0.1043 | -0.0685 |
| 4. Appearance | 0.1568[e.] | 0.0485 | 0.0697 |
| b. Community Services | -0.0352 | -0.0035 | 0.0580 |
| 1. Parks | -0.0102 | -0.0664 | 0.1026 |
| 2. Schools | -0.0525 | -0.1320[b.] | 0.0119 |
| 3. Security | -0.0128 | -0.0611 | 0.0168 |
| 4. Child Care | -0.0410 | -0.1333[b.] | -0.0034 |
| 5. Transportation | 0.0156 | -0.0045 | 0.0765 |
| 6. Entertainment | -0.0336 | -0.0898 | -0.0163 |
| c. Social Patterns | -0.0539 | -0.2499[e.] | -0.0709 |
| 1. Friendships | -0.0522 | -0.1101[b.] | -0.1306[b.] |
| 2. Group Activities | -0.0391 | -0.1894[e.] | -0.0070 |
| 3. Sense of Belonging | -0.0479 | -0.1255[b.] | -0.0496 |
| d. Psychological Well-Being | -0.1190[e.] | -0.1978[e.] | 0.0087 |
| 1. Fullness vs. Emptiness of Life | -0.0851[c.] | -0.1623[c.] | 0.0408 |
| 2. Social Respect | -0.0822[c.] | -0.1542[c.] | -0.0018 |
| 3. Personal Freedom | -0.0888[c.] | -0.0601 | 0.0626 |
| 4. Companionship | -0.0426 | -0.1395[c.] | -0.0838 |
| 5. Tranquility vs. Anxiety | -0.0651[b.] | -0.1363[c.] | -0.0436 |
| 6. Self Approving vs. Guilty | -0.0633[b.] | -0.0982 | 0.0479 |
| 7. Self Confidence | -0.1094[e.] | -0.2104[e.] | 0.0464 |
| 8. Elation vs. Depression | -0.1084[e.] | -0.2105[e.] | -0.0276 |

a. two-tailed t-test yields p>0.100 unless otherwise noted
b. t-test yields p<0.100
c. t-test yields p<0.050
d. t-test yields p<0.010
e. t-test yields p<0.005

confirmed expectations, especially those associated with psychological well-being in the U.S., and in Israel; such was not the case in the Netherlands. None of the relationships between a respondent's age and satisfaction with community services was significant, but housing satisfaction in the Netherlands and satisfaction with social patterns in Israel proved to be strong correlates of age. The positive relationship between age and housing satisfaction is almost entirely a function of aesthetic qualities ($p<0.005$) which were sufficient to influence the overall score on housing satisfaction ($p<0.005$).

While housing satisfaction was not a significant factor in either Israel or in the Netherlands, social patterns in Israel are as age-sensitive as are psychological factors. All but two of the specific indexes comprising the psychological well-being index were found to correlate negatively with a woman's age insofar as the American and Israeli samples are concerned. In Israel, this trend continues with unanimity in the dependent variable cluster of social patterns. There, satisfaction with group activities and overall satisfaction with social patterns provided resounding confirmation of expectations ($p<0.005$ for both). When combined with the fact that five of eight specific indexes of psychological well-being in the United States and six of eight in Israel all registered partial correlation coefficients ($p<0.050$) and the overall satisfaction with psychological well-being inversely related to age (so that $p<0.005$), it may be concluded that age is a powerful determinant of quality of life considerations.

Discussion: A woman who is older has had more time to settle in the sort of neighborhood she desires as determined by the interaction of individual preferences and the range of choices provided by the market. While this may be true in the United States, it does not generally hold in the Netherlands, where housing is an item of chronic scarcity, a factor which severely limits choice. Similarly, in Israel, a nation which has grown in population largely due to immigration, the government has been able to satisfy the need for shelter at the expense of the consumer sovereignty that is taken for granted in the United States.

Nevertheless, in the United States, there are factors which provide an explanation for the relationships suggested by the data. Women who are older apparently enjoy their housing environment more than younger women regardless of the length of time they have lived in a given dwelling unit. There is some evidence that expectations of the material world decrease with age. Perhaps as women age convenience and familiarity are preferred to opulence. Research has suggested that later in a woman's life cycle her relationship with her home surroundings takes on a high order of importance and that older, married women and widows enjoy housework more than other life cycle groups.[9]

The negative scores on the various indexes of psychological well-being are not surprising. It is an inescapable fact that we live in a youth-oriented culture and that women in particular have been socialized to have devalued the aging process. Nevertheless, despite the progress made by the women's movement, it was surprising to encounter the very strong association of age with lower satisfaction scores on nearly every specific index of psychological well-being among the respondents in the American and Israeli samples. One possible explanation which was considered was that one particular age group, perhaps the oldest women in the sample (which ranged in years from 20 to 50) had scored drastically lower than the other age groups. After dividing the sample into six age ranges and comparing the mean scores of these groups on dependent variables, one-way analysis of variance failed to reveal any significant differences among the various age groups. Therefore, it may be concluded that the relationships existing between a respondent's age and dependent variables operate consistently across the entire age range of the sample.

## 2. Marital Status

Expectation: Married women will have higher scores on all dependent satisfaction measures than will unmarried women.

Findings: This expectation is supported by the data but its strength varies among the three nations considered. However, in no instance were expectations refuted, while in every society at least one overall index of satisfaction produced significantly higher scores for married women.

As indicated in Table 4.15, housing satisfaction in the United States and in Israel is

TABLE 4.15 MEAN SCORES OF INDEXES OF SATISFACTION FOR MARITAL STATUS: U.S.A., ISRAEL AND THE NETHERLANDS

| Satisfaction Index | Satisfaction Scores[a.] | | | | | |
|---|---|---|---|---|---|---|
| | U.S.A. | | Israel | | Netherlands | |
| | Married (n=705) | Unmar. (n=120) | Married (n=284) | Unmar. (n=11) | Married (n=197) | Unmar. (n=18) |
| a. Housing Environment | 5.62[c.] | 5.40 | 4.80 | 4.75 | 5.14 | 4.91 |
| 1. Privacy | 5.84 | 5.72 | 5.38 | 5. 8 | 4.94 | 4.79 |
| 2. House & Lot | 5.58 | 5.49 | 5.08 | 5.53 | 5.50 | 5.28 |
| 3. Household Responsibilities | 5.31[e.] | 4.63 | 4.23[c.] | 3.25 | 4.91 | 4.41 |
| 4. Appearance | 5.77 | 5.75 | 4.56 | 4.65 | 5.19 | 4.86 |
| b. Community Services | 5.13[b.] | 4.98 | 4.09 | 4.11 | 5.15[b.] | 4.78 |
| 1. Parks | 4.96 | 5.03 | 4.46 | 5.25 | 5.40[b.] | 4.62 |
| 2. Schools | 5.18 | 5.15 | 4.84 | 4.86 | 5.78 | 5.72 |
| 3. Security | 5.20 | 5.12 | 5.13 | 5.81 | 4.37[b.] | 3.78 |
| 4. Child Care | 4.23[c.] | 3.92 | 3.81 | 3.13 | 4.30[c.] | 3.36 |
| 5. Transportation | 5.46[c.] | 5.18 | 3.70 | 3.50 | 5.98 | 6.28[b.] |
| 6. Entertainment | 5.50 | 5.40 | 2.89 | 4.22[c.] | 4.99 | 5.14 |
| c. Social Patterns | 5.39[e.] | 4.96 | 4.46[c.] | 3.24 | 4.74 | 4.31 |
| 1. Friendships | 5.68 | 5.53 | 4.81[b.] | 4.04 | 5.29 | 4.78 |
| 2. Group Activities | 5.37[e.] | 4.90 | 3.52 | 3.42 | 4.65 | 4.50 |
| 3. Sense of Belonging | 5.13[e.] | 4.51 | 5.12[c.] | 3.89 | 4.21 | 3.58 |

| | | | | | | |
|---|---|---|---|---|---|---|
| d. Psychological Well-Being | 6.97[c.] | 6.77 | 6.92 | 6.41 | 6.55 | 6.41 |
| 1. Fullness vs. Emptiness of Life | 7.16[e.] | 6.63 | 6.62[e.] | 5.10 | 6.47[b.] | 6.00 |
| 2. Social Respect | 7.50 | 7.38 | 7.56 | 7.30 | 6.81 | 6.50 |
| 3. Personal Freedom | 6.74 | 6.80 | 6.76 | 6.80 | 6.93 | 6.94 |
| 4. Companionship | 7.56[e.] | 6.34 | 7.40 | 7.22 | 7.03 | 7.06 |
| 5. Tranquility vs. Anxiety | 6.32[d.] | 5.93 | 6.57[d.] | 5.11 | 6.15[b.] | 5.56 |
| 6. Self Approving vs. Guilty | 6.74 | 6.63 | 7.04 | 7.13 | 6.19 | 6.17 |
| 7. Self Confidence | 7.10 | 6.99 | 6.67 | 6.44 | 6.20 | 6.39 |
| 8. Elation vs. Depression | 6.70[c.] | 6.40 | 6.59[e.] | 5.33 | 6.61 | 6.67 |

a. two-tailed t-test for all score differences yields $p>0.100$ unless otherwise noted
b. t-test yields $p<0.100$
c. t-test yields $p<0.050$
d. t-test yields $p<0.010$
e. t-test yields $p<0.005$

attributable to household responsibilities being less difficult for married women ($p<0.005$ in the U.S., $p<0.050$ in Israel), a factor dictated solely by income. Community services were more satisfying for married women in the United States and in the Netherlands, particularly for child care ($p<0.050$ for both). Social patterns were more rewarding in the United States ($p<0.005$) and in Israel ($p<0.050$) but no statistical differences were observed in the Dutch sample. The impetus for this result was a greater sense of belonging in the United States and in Israel ($p<0.005$, $p<0.050$, respectively) and greater satisfaction for married women with group activities in the United States ($p<0.005$). All three societies reported at least some greater satisfaction among married women with various aspects of psychological well-being with overall satisfaction being statistically significant in the United States ($p<0.050$); among Israeli married women three specific indexes registered at least at 0.010 levels of confidence.

Discussion: The above findings suggest that physical and social environments in contemporary suburban life are more supportive for married women than for single or divorced women. Clearly, economic factors provide the underpinnings for these distinctions, particularly with regard to the housing environment in the United States and in Israel; in the Netherlands, housing costs are subsidized for all but the upper one-third of the population and thus do not exert this pervasive influence on the Dutch sample.

The following data, taken from the American sample, illustrate the force of the economic imperative. Comparable data from Israel and the Netherlands will not be presented due to the very small sub-samples of unmarried women interviewed in those two nations.* The household income of unmarried women from all sources ranged from $14,000 to $16,000, approximately one-third less than that for married women. Eighty-two percent of unmarried women were employed as compared with 46% of married women. The former group tends to work at occupations commensurate with their education (1.06 on the

* Only 10 of 280 Israeli women were unmarried (3.6%) while only 18 of 215 Dutch women were unmarried (8.4%). In the American sample, 14.5% (120 out of 825) were unmarried.

occupation/education ratio) while married women are generally underemployed (0.76 on the occupation/ education ratio), figures which suggest that unmarried women are required to work to their full capacity in order to maintain the type of home included in the sample. Although only 9.1% of the American sample were renters, single women were much more likely to rent than were married women (22% vs. 7%). Thus, the financial problems of the unmarried women are well-documented.

Even so, one is struck by the differentials in the percentage of unmarried women within the three samples. Perhaps American couples are less financially constrained from separating than are those in the Netherlands and Israel. It would not be facetious to suggest that the unprecedented escalation of housing values during the 1970s in the San Francisco Bay Area made it financially possible for couples to liquidate the equity in their homes and buy separate dwelling units with the proceeds. In contrast, Israeli households, which registered the highest rate of female employment, about 66%, might not be able to sustain the economic hardship of divorce due to high housing costs (comparable to those in the United States) and a high rate of inflation exacerbating an already tenuous situation. In the Netherlands, the extremely tight housing market in the western part of Randstad Holland also forecloses to a large degree the fragmentation of households. The generally higher satisfaction scores in the American sample raise the question of whether marital relationships, because they are more easily abandoned, result in a better quality of life among those married couples who choose to remain together.

In the area of satisfaction with community services, married women in the United States and in the Netherlands enjoy a distinct advantage over unmarried women. In the United States, this is attributable to the greater resources available to married women in the areas of child care and transportation, services which are dependent almost entirely upon the private sector for their provision. In the Netherlands, child care emerges as the most statistically sensitive variable to marital status ($p<0.050$). Satisfaction with security and with parks is also higher among married women ($p<0.100$) and could be attributed to the presence of a male partner. As for the Israeli sample, satisfaction scores for community services were almost equally low for both married and unmarried women when compared to their counterparts in other more affluent countries. It is only in one area of significant satisfaction, entertainment ($p<0.050$), which favors unmarried women who apparently have more opportunities to get out of the house to experience entertainment and cultural activities.

Higher satisfaction scores for married women with the social patterns of metropolitan life in the United States and in Israel lend themselves to ready explanation. Again, however, because of the very small samples of unmarried women in Israel and in the Netherlands, only detailed results from the United States will be utilized. Table 4.16 indicates that married women have more casual acquaintances in their neighborhoods, more friends with which to chat, borrow from, count on for aid in emergencies, etc. However, when discussion of personal problems is necessary, both married and unmarried alike have similar numbers of neighbors with which to exchange ideas. When asked how many close friends reside in neighborhoods, in the San Jose area, within an hour of their home, and further away, both married and unmarried women reported comparable numbers.

Group participation scores reflected a similar pattern. Unmarried women were not as involved in informal groups, action groups or associations related to their children's activities. Beyond the neighborhood, however, marital status is not a determinant of the level of involvement in political groups, informal social organizations, or other organized groups with the exception of religious affiliations. The greater level of participation for married women reflects the support received at home that helps free up time that ordinarily would go to child care and household maintenance. It is also related to the longer commutes that single women endure, a fact which emerges when urban environmental influences are discussed later in this chapter. It should be stressed that unmarried women are not less congenial, but their opportunities to socialize are more constrained.

It appears that single women are able to fulfill their needs for friendship and belonging through relationships and group involvement outside their neighborhood and through one or two closer friendships proximate to their home. While being dissatisfied by their relative lack of casual neighborly contact, they suspect that they are not an essential part of the neighborhood social system. It would be interesting to learn if this dissatisfaction springs from a sense of alienation

TABLE 4.16 MEAN SCORES OF INDEXES OF SATISFACTION: MARITAL STATUS AND SELECTED FACTORS OF FRIENDSHIPS AND GROUP INVOLVEMENT: THE SAN JOSE AREA

| Variable | Mean Scores[a] | |
|---|---|---|
| | Married (n=705) | Unmarried (n=120) |
| Number of speaking acquaintances in neighborhood | 23.73[e] | 16.40 |
| Neighbors to discuss topics of general interest | 9.40[e] | 7.00 |
| Neighbors to borrow from | 4.44[c] | 3.45 |
| Neighbors to aid in crises | 4.85[b] | 3.61 |
| Neighbors to aid in projects | 2.59[c] | 1.88 |
| Neighbors to discuss personal problems | 1.59[c] | 1.56 |
| Close friends in neighborhood | 2.37 | 2.11 |
| Close friends in San Jose Area | 8.75 | 8.52 |
| Close friends 1 hour away | 5.33 | 4.45 |
| Close friends farther away | 13.60 | 16.61 |
| Satisfaction neighborhood friends | 5.67 | 5.49 |
| Neighborhood informal groups | 2.15[c] | 1.75 |
| Neighborhood action projects | 2.26[e] | 1.68 |
| Children's activity groups | 3.63[e] | 2.85 |
| Religious groups | 2.75[e] | 2.03 |
| Political organizations | 1.69 | 1.53 |
| Outside informal groups | 2.77 | 2.97 |
| Outside action groups | 1.86 | 1.84 |
| Outside social organization | 1.92 | 2.14 |
| Other organized groups | 2.00 | 2.10 |
| Satisfaction with group opportunities in neighborhood | 5.17[e] | 4.63 |
| Satisfaction with group opportunities in San Jose Area | 5.45 | 5.21 |
| Identification with neighborhood | 4.76[e] | 4.01 |
| Satisfaction with neighborhood identification | 5.49[e] | 5.01 |

a. two-tailed t-test for all score differences yields $p>0.100$ unless otherwise noted
b. t-test for score difference yields $p<0.100$
c. t-test for score difference yields $p<0.050$
d. t-test for score difference yields $p<0.010$
e. t-test for score difference yields $p<0.005$

from the stream of neighborhood socializing.

Although married women scored higher than unmarried women on various specific indexes of psychological well-being in all three samples, it is interesting that two specific indexes registered significant differences in all three samples: fullness vs. emptiness of life, tranquility vs. anxiety and elation vs. depression were significant in the United States and Israel, while companionship, interestingly enough, was significant only in the United States. Because these are, subjectively, the most global in their impact, marital status does appear to wield a substantial influence on psychological well-being.

### 3. Number and Age of Children

Expectation: All dependent satisfaction indexes will vary negatively with the number of children and positively with the age of children.

Findings: Survey data generally confirm that the number of children in a given household does vary negatively with indexes of satisfaction although only in the Israeli sample did this expectation register statistical significance in all four dependent variable clusters (see Table 4.17). In the United States support for this contention was limited to only the housing environment wherein overall satisfaction plus that on three of four specific indexes registered $p<0.005$. In the Netherlands, the number of children was inversely related to satisfaction scores only in the social patterns specific index of satisfaction with group activities ($p<0.050$), and one psychological index, freedom, was somewhat significant ($p<0.100$).

The broad-based support for expectations in Israel was most striking in the areas of psychological well-being and in social patterns (overall satisfaction in both $p<0.005$). With respect to the former, satisfaction with friendships and with group activities were strongly and inversely related to the number of children ($p<0.005$ for both). As for the latter, five of eight specific indexes strongly followed our expectation ($p<0.050$). The inverse relationship between the number of children and satisfaction with community services and housing was based on only two specific indexes. Satisfaction with entertainment ($p<0.005$) was responsible for the significant overall satisfaction score for community services, and satisfaction with

TABLE 4.17 RELATIONSHIPS BETWEEN NUMBER OF CHILDREN AND INDEXES OF SATISFACTION: U.S.A., ISRAEL AND THE NETHERLANDS

| Satisfaction Index | Partial Correlation Coefficient[a] | | |
|---|---|---|---|
| | U.S.A. (n=825) | Israel (n=295) | Netherlands (n=215) |
| a. Housing Environment | -0.1522[e] | -0.1575[c] | -0.0001 |
| 1. Privacy | -0.1143[e] | -0.0979 | 0.0783 |
| 2. House & Lot | -0.1117[e] | -0.0418 | 0.0268 |
| 3. Household Responsibilities | -0.1125[e] | -0.3122[e] | 0.0117 |
| 4. Appearance | -0.0788[c] | 0.0256 | -0.1010 |
| b. Community Services | -0.0190 | -0.1352[c] | -0.0321 |
| 1. Parks | -0.0400 | -0.0194 | -0.0447 |
| 2. Schools | 0.0285 | -0.0178 | 0.0051 |
| 3. Security | 0.0027 | -0.0806 | -0.0657 |
| 4. Child Care | -0.0138 | -0.1052 | -0.1161 |
| 5. Transportation | -0.0566 | -0.0359 | 0.0703 |
| 6. Entertainment | -0.0456 | -0.2130[e] | 0.0138 |
| c. Social Patterns | 0.0018 | -0.3457[e] | -0.0376 |
| 1. Friendships | 0.0018 | -0.2279[e] | 0.0881 |
| 2. Group Activities | -0.0057 | -0.2748[e] | -0.1735[c] |
| 3. Sense of Belonging | -0.0456 | -0.0958 | 0.0210 |
| d. Psychological Well-Being | 0.0097 | -0.2202[e] | -0.0875 |
| 1. Fullness vs. Emptiness of Life | 0.0426 | -0.1793[d] | -0.0657 |
| 2. Social Respect | -0.0172 | -0.0337 | 0.0221 |
| 3. Personal Freedom | -0.0109 | -0.1089[b] | -0.1424[b] |
| 4. Companionship | 0.0188 | -0.1589[c] | -0.0139 |
| 5. Tranquility vs. Anxiety | -0.0225 | -0.1613[c] | -0.0377 |
| 6. Self Approving vs. Guilty | 0.0014 | -0.0633 | -0.0715 |
| 7. Self Confidence | -0.0052 | -0.1626[c] | -0.0633 |
| 8. Elation vs. Depression | 0.0488 | -0.2102[e] | -0.0509 |

a. two-tailed t-test yields $p>0.100$ unless otherwise noted
b. t-test yields $p<0.100$
c. t-test yields $p<0.050$
d. t-test yields $p<0.010$
e. t-test yields $p<0.005$

household responsibilities ($p<0.005$) accounted entirely for overall housing environment satisfaction score.

Concerning the average age of children, the predicted relationship was not strongly supported by the data (see Table 4.18) although some conformance to expectations was registered in the United States with overall satisfaction with psychological well-being ($p<0.050$) and in Israel with satisfaction with parks ($p<0.005$). The American sample expressed overall satisfaction with psychological well-being based on the positive relationship of children's age and satisfaction with personal freedom ($p<0.005$). Unexpected links between this expectation and the data were found in all three societies. In the United States and in Israel, housing satisfaction did not increase as children got older; to the contrary, satisfaction decreased, especially with respect to household responsibilities ($p<0.050$ in Israel, $p.<0.100$ in the U.S.) and appearance ($p<0.050$ in the U.S.). No statistical relationship occurred in the Netherlands with respect to housing satisfaction. There, however, satisfaction with parks was inversely related to children's age ($p<0.050$), thus again refuting expectations.

<u>Discussion</u>: The failure of the variables of number and age of children to predict as many dimensions of satisfaction as had been expected might be explained by the generally small size of the families in the sample. In line with a general trend among middle-class households, the participants in this survey averaged 2.12 children in the United States, 2.32 in Israel and 2.05 in the Netherlands. Thus, the range in family size may not have been great enough to have reflected the full impact of this variable.

The average age of children correlates in the American sample with certain aspects of psychological well-being, particularly personal freedom. In the Netherlands, no such correlations were observed, but in Israel, the relationship with the companionship specific index was negative, as was satisfaction with household responsibilities. This suggests that the physical ambience of the dwelling unit creates problems which spill over into the psychological realm. However, while the dwelling unit may pose some problems, satisfaction with public services, especially parks ($p<0.005$), does increase with a child's age. In contrast, the Dutch sample was not as fortunate with satisfaction with

TABLE 4.18 RELATIONSHIPS BETWEEN AVERAGE AGE OF CHILDREN AND INDEXES OF SATISFACTION: U.S.A., ISRAEL AND THE NETHERLANDS

| Satisfaction Index | Partial Correlation Coefficient[a] | | |
|---|---|---|---|
| | U.S.A. (n=825) | Israel (n=295) | Netherlands (n=215) |
| a. Housing Environment | -0.0218 | -0.0156 | 0.0353 |
| 1. Privacy | 0.0304 | 0.0491 | 0.0235 |
| 2. House & Lot | 0.0476 | 0.0291 | 0.0140 |
| 3. Household Responsibilities | -0.0630[b] | -0.1525[c] | 0.0085 |
| 4. Appearance | -0.0754[c] | 0.0099 | -0.0126 |
| b. Community Services | -0.0250 | 0.0634 | -0.1563[b] |
| 1. Parks | 0.0181 | 0.2022[e] | -0.1844[c] |
| 2. Schools | 0.0203 | 0.0941 | -0.0479 |
| 3. Security | 0.0340 | 0.1086 | 0.1021 |
| 4. Child Care | -0.0527 | 0.0143 | -0.0850 |
| 5. Transportation | -0.0406 | 0.0220 | -0.0438 |
| 6. Entertainment | 0.0315 | 0.1144[b] | 0.0066 |
| c. Social Patterns | 0.0053 | -0.0047 | 0.0483 |
| 1. Friendships | 0.0209 | 0.0082 | 0.0967 |
| 2. Group Activities | 0.0364 | -0.0280 | 0.0184 |
| 3. Sense of Belonging | -0.0218 | 0.0347 | -0.0096 |
| d. Psychological Well-Being | 0.0817[c] | 0.0438 | 0.0153 |
| 1. Fullness vs. Emptiness of Life | 0.0654[b] | -0.0120 | 0.0621 |
| 2. Social Respect | 0.0649[b] | 0.0701 | 0.0196 |
| 3. Personal Freedom | 0.1036[e] | -0.0334 | 0.0107 |
| 4. Companionship | 0.0519 | -0.1970[e] | 0.0128 |
| 5. Tranquility vs. Anxiety | 0.0125 | -0.0358 | -0.0089 |
| 6. Self Approving vs. Guilty | 0.0659[b] | 0.0122 | -0.0005 |
| 7. Self Confidence | 0.0327 | 0.0380 | -0.0750 |
| 8. Elation vs. Depression | 0.0591[b] | -0.0418 | 0.0656 |

a. two-tailed t-test yields $p>0.100$ unless otherwise noted
b. t-test yields $p<0.100$
c. t-test yields $p<0.050$
d. t-test yields $p<0.010$
e. t-test yields $p<0.005$

parks ($p<0.050$), a factor possibly due to the more compact metropolitan areas found in the Netherlands and the ensuing shortage of open space, even in a society which has a richer public sector than the other two studied here.

Further, recognizing the potential problems relating to the use of the average age of children as an indicator of the various factors in a mother's life, the United States data was recoded to account for the age of the youngest child. Only one index of satisfaction, that with schools, showed a significant relationship with age of the youngest child and it was inverse ($p<0.050$). Surprisingly, an analysis of variance did not reveal any statistical differences between respondents whose youngest child was of preschool age and those whose youngest child fell into other age brackets.

This secondary analysis revealed one curious characteristic of the sample. While the mean age of the youngest child in the household was 6.8 years, the modal value for youngest child was one year. Seventy-nine respondents, all mothers of one-year old children, participated in the study, but it is doubtful that this nearly 10% sample is characteristic of the population as a whole. No doubt it reflects that being tied to the home is a major characteristic of mothers of young children and, in addition, these individuals might be more willing than others to grant an interview in order to be in the company of another adult.

In summary, then, the failure of this analysis to discover any major differences in satisfaction scores based on the age of the youngest child refutes the notion of the isolated and unhappy mother of such an individual. Thus, it appears that the age and number of children of a respondent will not reveal as much about her general satisfaction with the quality of life as had been expected, but it will provide more information than knowledge of the age of the youngest child.

## D. SUBURBAN ENVIRONMENTAL INFLUENCES

### 1. Density

Expectation: Density will have negative relationships with dependent satisfaction variables concerned with the housing environment and social patterns for married women and the reverse relationships for women who are single heads of households.

Findings: Because of the scarcity of middle income low density neighborhoods of single family houses in Israel and in the western Netherlands, the sample was limited to a very narrow range of relatively high density housing in these countries (from about 30 to 35 housing units per net residential acre). Accordingly, we have confined our examination of the density variable only to the American study (from approximately 6 to 20 housing units per net residential acre).

Our data partially supports our expectation in the United States. Using our sample of married women and comparing multi-family units with single-family areas, Table 4.19 indicates that density has unexpected positive relationships with the overall satisfaction of the housing environment ($p<0.005$) and community services ($p<0.005$). The positive relationship between density and overall satisfaction of the housing environment centered around the relatively high satisfaction with household responsibilities (time and money) that residents of high density areas expressed. The desirability of density with regard to community services was based on the higher satisfaction expressed for the site and management characteristics of condominiums (park, security, and cultural and entertainment activities) compared to that for single-family areas.

The inconclusive finding about the relationship of density and social patterns for married women was the result of two offsetting superiorities: a higher satisfaction of group participation among high density residents; and a greater sense of belonging for single family residents. As expected, the scores of psychological well-being were inconclusive.

With regard to unmarried women, Table 4.20 shows that the data yielded the expected positive relationship between density and the overall satisfaction of the housing environment ($p<0.050$) as well as with community services ($p<0.100$). Also, similar to the findings for married women, the data show inconclusive relationships between density and the overall satisfaction of social patterns and psychological well-being. Not only were findings for unmarried women about the relationships between density and overall satisfaction measures similar to those for married women, but the underlying factors for these relationships were also similar. For example, it appears that the primary reason for the positive linkage between density and the housing

TABLE 4.19 MEAN SCORES OF INDEXES OF SATISFACTION: MARRIED WOMEN IN MULTI-FAMILY AND SINGLE-FAMILY HOUSING IN THE SAN JOSE AREA

| Satisfaction Index | Satisfaction Scores[a] | |
|---|---|---|
| | Multi-Family (n=231) | Single-Family (n=471) |
| a. Housing Environment | 5.79[e] | 5.54 |
| 1. Privacy | 5.81 | 5.85 |
| 2. House & Lot | 5.55 | 5.59 |
| 3. Household Responsibilities | 5.99[e] | 4.99 |
| 4. Appearance | 5.80 | 5.75 |
| b. Community Services | 5.28[e] | 5.05 |
| 1. Parks | 5.11[c] | 4.89 |
| 2. Schools | 5.17 | 5.19 |
| 3. Security | 5.69[e] | 5.05 |
| 4. Child Care | 4.30 | 4.19 |
| 5. Transportation | 5.42 | 5.47 |
| 6. Entertainment | 5.75[e] | 5.38 |
| c. Social Patterns | 5.37 | 5.40 |
| 1. Friendships | 5.69 | 5.68 |
| 2. Group Activities | 5.50[e] | 5.23 |
| 3. Sense of Belonging | 4.91 | 5.23[d] |
| d. Psychological Well-Being | 7.00 | 6.96 |
| 1. Fullness vs. Emptiness of Life | 7.12 | 7.18 |
| 2. Social Respect | 7.46 | 7.52 |
| 3. Personal Freedom | 6.85 | 6.70 |
| 4. Companionship | 7.57 | 7.56 |
| 5. Tranquility vs. Anxiety | 6.40 | 6.28 |
| 6. Self Approving vs. Guilty | 6.75 | 6.74 |
| 7. Self Confidence | 7.09 | 7.12 |
| 8. Elation vs. Depression | 6.69 | 6.70 |

a. two-tailed t-test yields $p>0.100$ unless otherwise noted
b. t-test yields $p<0.100$
c. t-test yields $p<0.050$
d. t-test yields $p<0.010$
e. t-test yields $p<0.005$

TABLE 4.20 MEAN SCORES OF INDEXES OF SATISFACTION: UNMARRIED WOMEN IN MULTI-FAMILY AND SINGLE-FAMILY HOUSING IN THE SAN JOSE AREA

| Satisfaction Index | Satisfaction Scores[a] | |
|---|---|---|
| | Multi-Family (n=78) | Single-Family (n=43) |
| a. Housing Environment | 5.52[c] | 5.12 |
| 1. Privacy | 5.74 | 5.74 |
| 2. House & Lot | 5.42 | 5.60 |
| 3. Household Responsibilities | 5.05[e] | 3.78 |
| 4. Appearance | 5.87 | 5.49 |
| b. Community Services | 5.07[e] | 4.77 |
| 1. Parks | 5.06 | 4.95 |
| 2. Schools | 5.26 | 4.91 |
| 3. Security | 5.29[c] | 4.81 |
| 4. Child Care | 3.89 | 4.01 |
| 5. Transportation | 5.27 | 5.00 |
| 6. Entertainment | 5.45 | 5.31 |
| c. Social Patterns | 4.94 | 4.95 |
| 1. Friendships | 5.40 | 5.66 |
| 2. Group Activities | 5.00 | 4.65 |
| 3. Sense of Belonging | 4.45 | 4.65 |
| d. Psychological Well-Being | 7.38 | 6.68 |
| 1. Fullness vs. Emptiness of Life | 7.14 | 6.65 |
| 2. Social Respect | 8.96 | 7.42 |
| 3. Personal Freedom | 8.38 | 6.71 |
| 4. Companionship | 7.49 | 7.24 |
| 5. Tranquilty vs. Anxiety | 6.13 | 5.70 |
| 6. Self Approving vs. Guilty | 6.91 | 6.67 |
| 7. Self Confidence | 7.50 | 7.05 |
| 8. Elation vs. Depression | 6.69 | 6.26 |

a. two-tailed t-test yields $p>0.100$ unless otherwise noted
b. t-test yields $p<0.100$
c. t-test yields $p<0.050$
d. t-test yields $p<0.010$
e. t-test yields $p<0.005$

TABLE 4.21 MEAN SCORES OF OVERALL INDEXES OF SATISFACTION: EMPLOYMENT STATUS AND DENSITY IN THE SAN JOSE AREA

| Employment | Overall Indices of Satisfaction[a] | Density | |
|---|---|---|---|
| | | Multi-Family Housing (n=306) | Single-Family Housing (n=513) |
| Non-Working | Housing Environment | 5.79[c] | 5.59 |
| | Community Services | 5.26[c] | 5.04 |
| | Social Patterns | 5.36 | 5.35 |
| | Psychological Well-Being | 6.98 | 6.89 |
| (n=399) | | (n=111) | (n=288) |
| Working | Housing Environment | 5.69[e] | 5.40 |
| | Community Services | 5.21[d] | 5.01 |
| | Social Patterns | 5.22 | 5.38 |
| | Psychological Well-Being | 7.17 | 6.98 |
| (n=420) | | (n=195) | (n=255) |

a. two-tailed t-test for all score differences yields $p>0.100$ unless otherwise noted
b. t-test for score difference yields $p<0.100$
c. t-test for score difference yields $p<0.050$
d. t-test for score difference yields $p<0.010$
e. t-test for score difference yields $p<0.005$

environment was also the relatively easier time women living in high density housing had dealing with household responsibilities.

Discussion: It appears that higher density housing areas not only work better for unmarried women than do single family neighborhoods, but are also more rewarding environments for married women with small children.

If the analysis is extended to consider the relationships of employment status and density, a pattern prevails which is similar to that found for marital status. As Table 4.21 reveals, the overall satisfaction scores for the housing environment and community services were significantly higher for both working ($p<0.010$) and non-working ($p<0.050$) women living in multi-family units compared to their counterparts in single-family housing. And these differences were based on specific indexes of satisfaction similar to those found for marital status.

Finally, if we examine the relationships of density and age of housing, it is found once again that the desirability of the multi-family housing units emerges with respect to the housing environment ($p<0.050$) and community services ($p<0.100$).[10] Thus, our findings strongly suggest that multi-family areas provide more efficient and better organized housing environments and a more supportive set of community services for all women with children than do single-family housing areas.

2. Household Distance from the Central City Center

Expectations: Household distance from the central city center (miles) varies negatively with the satisfaction variables concerned with community services, social patterns, and psychological well-being.

Findings: This expectation is supported strongly by the American data. Comparing an inlying single family neighborhood with an outlying single family area, Table 4.22 indicates a negative relationship between household distance from the central city center and all overall indexes of satisfaction ($p<0.050$). As anticipated, the satisfaction of community services was especially high for the inlying residents ($p<0.005$) and for transportation services in particular ($p<0.005$) due to their relatively greater proximity to the city center.

TABLE 4.22 MEAN SCORES OF INDEXES OF SATISFACTION: INLYING AND OUTLYING SINGLE-FAMILY HOUSING NEIGHBORHOODS IN SAN JOSE

| Satisfaction Index | Satisfaction Scores[a] | |
|---|---|---|
| | Inlying Area (n=103) | Outlying Area (n=100) |
| a. Housing Environment | 5.91[e] | 5.32 |
| 1. Privacy | 6.21 | 5.66 |
| 2. House & Lot | 6.02[e] | 5.41 |
| 3. Household Responsibilities | 4.90 | 4.82 |
| 4. Appearance | 6.52[e] | 5.32 |
| b. Community Services | 5.07[e] | 4.45 |
| 1. Parks | 5.60[e] | 3.71 |
| 2. Schools | 4.75 | 4.83 |
| 3. Security | 5.08[c] | 4.55 |
| 4. Child Care | 4.42[c] | 4.04 |
| 5. Transportation | 5.47[e] | 4.83 |
| 6. Entertainment | 5.06 | 4.98 |
| c. Social Patterns | 5.67[c] | 5.33 |
| 1. Friendships | 5.92 | 5.72 |
| 2. Group Activities | 5.41 | 5.11 |
| 3. Sense of Belonging | 5.60[c] | 5.07 |
| d. Psychological Well-Being | 7.13[c] | 6.84 |
| 1. Fullness vs. Emptiness of Life | 7.34[c] | 6.94 |
| 2. Social Respect | 7.58 | 7.55 |
| 3. Personal Freedom | 6.69 | 6.48 |
| 4. Companionship | 7.90[c] | 7.51 |
| 5. Tranquilty vs. Anxiety | 6.49[c] | 6.01 |
| 6. Self Approving vs. Guilty | 6.93 | 6.72 |
| 7. Self Confidence | 7.18 | 7.05 |
| 8. Elation vs. Depression | 6.89[b] | 6.56 |

a. two-tailed t-test yields $p>0.100$ unless otherwise noted
b. t-test yields $p<0.100$
c. t-test yields $p<0.050$
d. t-test yields $p<0.010$
e. t-test yields $p<0.005$

Other specific indexes of satisfaction which as expected, indicated significantly higher scores ($p<0.050$) for the inlying area are sense of belonging, fullness of life, companionship, and tranquility. Unanticipated was the relatively high satisfaction with the housing environment reported by inlying area residents based primarily on the design aspects of the house and lot, and appearance (all $p<0.005$).

In contrast, our findings comparing inlying and outlying areas in Israel and the Netherlands did not support expectations. As Table 4.23 shows, with the exception of the desirability of inlying Israeli neighborhoods' accessibility to entertainment facilities ($p<0.005$), outlying Israeli areas scored high in satisfaction with the quality of community services, such as parks ($p<0.050$) and especially in the realm of psychological well-being ($p<0.005$). The Netherlands data shown on Table 4.24 also indicate a preference for outlying areas for such community services as parks, and the quality of friendships ($p<0.005$).

<u>Discussion</u>: It is possible that some of the satisfaction expressed by residents in the inlying American area was due to the overall aesthetic qualities of the area, since it is an older established tree-lined residential neighborhood with a strong physical and social identity compared to the newer outlying area. Yet, when the sample is controlled for age of housing, the across-the-board superiority of the inlying neighborhood prevailed, with the exception of psychological well-being scores.[11]

Thus, despite the highly decentralized pattern of shopping, entertainment, and cultural facilities in the San Jose area, findings suggest that the accessibility to the city center available to inlying neighborhood residents provides them with many physical and social benefits which residents in outlying areas do not receive. These results corroborate the findings of others that women living in outlying American suburban areas experience difficulties related to accessibility for such activities as employment[12] and when not socially involved locally, residents of these outlying areas tend to be more isolated than their central city counterparts.[13] Perhaps, as a metropolitan area grows geographically and becomes more suburban in character, its population becomes increasingly localized and regionally isolated in nature.

TABLE 4.23 MEAN SCORES OF INDEXES OF SATISFACTION: INLYING AND OUTLYING HOUSING NEIGHBORHOODS IN ISRAEL

| Satisfaction Index | Satisfaction Scores[a] | |
|---|---|---|
| | Inlying Areas (n=110) | Outlying Areas (n=185) |
| a. Housing Environment | 4.77 | 4.81 |
| 1. Privacy | 5.35 | 5.40 |
| 2. House & Lot | 4.95 | 5.15 |
| 3. Household Responsibilities | 4.36 | 4.13 |
| 4. Appearance | 4.54 | 4.58 |
| b. Community Services | 3.84 | 4.18 |
| 1. Parks | 4.08 | 4.62[c] |
| 2. Schools | 4.84 | 4.82 |
| 3. Security | 4.98 | 5.21 |
| 4. Child Care | 3.44 | 3.90[b] |
| 5. Transportation | 3.56 | 3.76 |
| 6. Entertainment | 3.77[e] | 2.60 |
| c. Social Patterns | 4.41 | 4.43 |
| 1. Friendships | 4.73 | 4.80 |
| 2. Group Activities | 3.69 | 3.45 |
| 3. Sense of Belonging | 4.93 | 5.12 |
| d. Psychological Well-Being | 6.62 | 7.03[e] |
| 1. Fullness vs. Emptiness of Life | 6.55 | 6.60 |
| 2. Social Respect | 8.15 | 7.68 |
| 3. Personal Freedom | 6.34 | 6.93[d] |
| 4. Companionship | 7.13 | 7.61[c] |
| 5. Tranquilty vs. Anxiety | 6.11 | 6.69[e] |
| 6. Self Approving vs. Guilty | 6.80 | 7.15[c] |
| 7. Self Confidence | 6.32 | 6.78[d] |
| 8. Elation vs. Depression | 6.34 | 6.64[b] |

a. two-tailed t-test yields $p>0.100$ unless otherwise noted
b. t-test yields $p<0.100$
c. t-test yields $p<0.050$
d. t-test yields $p<0.010$
e. t-test yields $p<0.005$

TABLE 4.24 MEAN SCORES OF INDEXES OF SATISFACTION: INLYING AND OUTLYING HOUSING NEIGHBORHOODS IN THE NETHERLANDS

| Satisfaction Index | Satisfaction Scores[a] | |
|---|---|---|
| | Inlying Areas (n=105) | Outlying Areas (n=110) |
| a. Housing Environment | 5.14 | 5.11 |
| 1. Privacy | 4.96 | 4.88 |
| 2. House & Lot | 5.45 | 5.52 |
| 3. Household Responsibilities | 5.02[b] | 4.73 |
| 4. Appearance | 5.10 | 5.23 |
| b. Community Services | 5.10 | 5.13 |
| 1. Parks | 5.09 | 5.58[e] |
| 2. Schools | 5.82 | 5.76 |
| 3. Security | 4.40 | 4.24 |
| 4. Child Care | 4.35 | 4.13 |
| 5. Transportation | 6.04 | 5.98 |
| 6. Entertainment | 4.90 | 5.13 |
| c. Social Patterns | 4.61 | 4.81 |
| 1. Friendships | 4.96 | 5.54[e] |
| 2. Group Activities | 4.64 | 4.66 |
| 3. Sense of Belonging | 4.16 | 4.14 |
| d. Psychological Well-Being | 6.59 | 6.48 |
| 1. Fullness vs. Emptiness of Life | 6.48 | 6.38 |
| 2. Social Respect | 6.83 | 6.72 |
| 3. Personal Freedom | 7.10[b] | 6.78 |
| 4. Companionship | 7.10 | 6.96 |
| 5. Tranquilty vs. Anxiety | 6.07 | 6.11 |
| 6. Self Approving vs. Guilty | 6.27 | 6.12 |
| 7. Self Confidence | 6.26 | 6.17 |
| 8. Elation vs. Depression | 6.65 | 6.58 |

a. two-tailed t-test for all score differences yields $p>0.100$ unless otherwise noted

b. t-test yields $p<0.100$

c. t-test yields $p<0.050$

d. t-test yields $p<0.010$

e. t-test yields $p<0.005$

While there were a few benefits of Israeli and Dutch inlying neighborhoods, the outlying areas in these countries were found to be superior, even when controlling for age of housing, density, and some of the social characteristics of the residents such as marital and employment status.[14] It appears that other forces were at work to generate suburban preferences in these nations. Perhaps the attractive outlying suburban setting, previously denied to most middle class households in Israel and the Netherlands, has become highly prized not only because it offers real physical and social rewards, but also because it symbolizes improved living standards in these nations, as was the case in the United States during the 1950s and 1960s.

### 3. Age of Neighborhood

Expectation: Age of neighborhood will have a positive association with all dependent satisfaction variables.

Findings: The findings generally support this expectation for the United States. As Table 4.25 indicates, a partial correlation analysis shows a positive relationship between age of neighborhood and the overall satisfaction with three aspects of the housing environment ($p<0.005$), social patterns ($p<0.050$) and two indexes of psychological well-being ($p<0.050$). With the sole exception of the negative relationship with satisfaction of household responsibilities ($p<0.005$), age of neighborhood conformed to expectations of the housing environment at the 0.005 level.

The social success of the older housing environments in San Jose was based primarily on the positive relationships between age of neighborhood and satisfaction with friendships ($p<0.050$) and sense of belonging ($p<0.005$). Also the positive linkage of the older neighborhoods to psychological well-being was primarily socially oriented as indicated by the positive relationships between the age of neighborhood and social respect and companionship at the 0.050 level. Apart from the positive relationship with satisfaction of parks ($p<0.005$) the age of neighborhood had no significant linkages with community services.

Israeli and Dutch findings generally do not support the expectation concerning age of neighborhood. In no instance did this variable have a positive significant relationship on satisfaction

TABLE 4.25 RELATIONSHIPS BETWEEN AGE OF NEIGHBORHOOD AND INDEXES OF SATISFACTION: U.S.A., ISRAEL AND THE NETHERLANDS

| Satisfaction Index | Partial Correlation Coefficient[a.] | | |
|---|---|---|---|
| | U.S.A. (n=825) | Israel (n=295) | Netherlands (n=215) |
| a. Housing Environment | 0.0594[b.] | 0.0073 | -0.1695[c.] |
| 1. Privacy | 0.1076[e.] | 0.0753 | -0.1959[c.] |
| 2. House & Lot | 0.1127[e.] | -0.0117 | -0.0212 |
| 3. Household Responsibilities | -0.1990[e.] | -0.0627 | -0.1939[c.] |
| 4. Appearance | 0.1718[e.] | 0.0344 | -0.0113 |
| b. Community Services | 0.0185 | 0.1120 | -0.0380 |
| 1. Parks | 0.1325[e.] | 0.0563 | -0.0975 |
| 2. Schools | 0.0008 | 0.1278 | -0.0054 |
| 3. Security | -0.0128 | 0.0789 | -0.0049 |
| 4. Child Care | -0.0111 | -0.0204 | -0.0235 |
| 5. Transportation | -0.0129 | -0.0057 | -0.0617 |
| 6. Entertainment | -0.0402 | 0.0915 | -0.0262 |
| c. Social Patterns | 0.0841[c.] | -0.0745 | -0.0367 |
| 1. Friendships | 0.0760[c.] | -0.1610[c.] | -0.0619 |
| 2. Group Activities | 0.0140 | -0.0056 | -0.1351[b.] |
| 3. Sense of Belonging | 0.1034[e.] | 0.0027 | 0.0640 |
| d. Psychological Well-Being | 0.0571[b] | -0.1187[b.] | -0.1132 |
| 1. Fullness vs. Emptiness of Life | 0.0634[b.] | -0.0370 | -0.0609 |
| 2. Social Respect | 0.0711[c.] | -0.0778 | -0.0178 |
| 3. Personal Freedom | -0.0587[b.] | 0.0482 | -0.0580 |
| 4. Companionship | 0.0769[c.] | -0.2230e. | -0.1063 |
| 5. Tranquility vs. Anxiety | 0.0295 | -0.0835 | -0.0819 |
| 6. Self Approving vs. Guilty | 0.0598[b.] | -0.1462 | -0.0155 |
| 7. Self Confidence | 0.0559[b.] | -0.0169 | -0.0638 |
| 8. Elation vs. Depression | 0.0462 | -0.0001 | -0.1573[c.] |

a. two-tailed t-test yields p>0.100 unless otherwise noted
b. t-test yields p<0.100
c. t-test yields p<0.050
d. t-test yields p<0.010
e. t-test yields p<0.005

in both countries. In Israel, age of neighborhood had its primary negative relationships with specific indexes of satisfaction in the social and psychological realms of friendships ($p<0.050$) and companionship ($p<0.005$). In the Netherlands, the negative relationship of age of neighborhood was focussed predominantly on the overall satisfaction of housing environment ($p<0.050$) especially the specific indexes of privacy and household responsibilities.

Discussion: While the older neighborhoods in the United States seem to have the disadvantage of housing with a high incidence of maintenance responsibilities, these housing environments are still considered more desirable than the newer areas because of the satisfaction they create in privacy, arrangement and design of house and lot, and overall appearance. More importantly, the older American neighborhoods seem to provide a more cohesive, stable and supportive social and psychological network for their residents than do new residential areas. Apparently, people are willing to tolerate the annoying maintenance costs of older housing, and possible social constraints in order to receive the aesthetic, social, and psychological benefits of living in established neighborhoods. It seems that the older residential areas with their tree-lined streets, traditional housing appearance, and varied age population structure create the image and feeling of continuity in an era of rapid change and great household mobility in the United States.

In many ways the reverse may be true for Israel and the Netherlands. Despite the long-standing preference for established older neighborhoods by Israeli middle class families for reasons of central city accessibility, newer study areas were found to be more socially and psychologically rewarding.[15] What may be happening in Israel is that residents of these newer and often more socially heterogeneous suburban areas are experiencing a broader and richer social and psychological network than their old neighborhood counterparts.[16] This may especially be the case for Jewish families from the Middle East and North Africa who were often ghettoized in old central city Israeli neighborhoods.[17]

The superiority of the new Dutch neighborhoods seems to be largely attributable to a real change in housing quality. That is, since the early 1960s, the Netherlands has been able to produce steadily improved housing quality aimed specifically at the middle income housing market.[18] Thus, recently developed Dutch neighborhoods tend to be more desirable than older areas because of the demand for the superior amenities found in ncw housing units.

## 4. Distance to Work

Expectation: Travel time to work of a woman and/or her spouse will vary negatively with her satisfaction with community services, social patterns and psychological well-being.

Findings: This expectation was partially supported by our data from the United States. The correlations shown on Table 4.26 indicate that women's travel time to work varies negatively with only one overall index of satisfaction, social patterns ($p<0.050$), mostly because of the apparently detrimental impacts of long journeys to work on the satisfaction of friendships and group activities ($p<0.050$). With the exception of the negative linkages to the specific index of psychological well-being of fullness of life ($p<0.050$), women's travel time to work had no other significant relationships with measures of women's satisfaction in the United States.

In the Netherlands, expectations were also only partially confirmed. Here, women's travel time to work varied negatively with the specific indexes of child care ($p<0.050$) and friendships ($p<0.100$) and had a positive linkage to self-confidence ($p<0.100$).

Israel, however, generated data which strongly supported expectations. The Israeli women's travel time to work varied negatively with the overall satisfaction of housing environment ($p<0.050$), community services ($p<0.005$) and social patterns ($p<0.100$), as well as with the specific psychological index of social respect ($p<0.100$).

More in keeping with expectations were the negative relationshps that American men's travel time to work has with the overall indexes of satisfaction of community services ($p<0.050$) and social patterns ($p<0.100$), as shown in Table 4.27. The dissatisfaction with community services was related to the negative scores for transportation ($p<0.050$) and entertainment and culture ($p<0.100$), while the social dissatisfaction was based on the negative value for sense of belonging ($p<0.100$). The slight dissatisfactions of psychological well-being were similar to those found with respect to women's travel time. Unexpectedly, men's travel time to work in the United States was also negatively associated

TABLE 4.26 RELATIONSHIPS BETWEEN WOMAN'S TRAVEL TIME TO WORK AND INDEXES OF SATISFACTION: U.S.A., ISRAEL AND THE NETHERLANDS

| Satisfaction Index | Partial Correlation Coefficient[a] | | |
|---|---|---|---|
| | U.S.A. (n=825) | Israel (n=295) | Netherlands (n=215) |
| a. Housing Environment | -0.0054 | -0.1589[c] | -0.0433 |
| 1. Privacy | -0.0012 | -0.1180[b] | -0.0176 |
| 2. House & Lot | -0.0235 | -0.1091[b] | 0.0258 |
| 3. Household Responsibilities | -0.0023 | -0.1492[c] | -0.0790 |
| 4. Appearance | 0.0082 | -0.0523 | -0.0198 |
| b. Community Services | 0.0220 | -0.3025[e] | -0.0726 |
| 1. Parks | -0.0272 | -0.1005 | 0.0034 |
| 2. Schools | 0.0514 | -0.1115 | -0.0863 |
| 3. Security | -0.0015 | -0.1844[d] | -0.1085 |
| 4. Child Care | -0.0218 | -0.1073 | -0.1627[c] |
| 5. Transportation | -0.0327 | -0.1296[b] | 0.1174 |
| 6. Entertainment | 0.0101 | -0.0902 | 0.0726 |
| c. Social Patterns | -0.0795[c] | -0.1246[b] | -0.1242 |
| 1. Friendships | -0.0720 | 0.0052 | -0.1546[b] |
| 2. Group Activities | -0.0803[c] | -0.0879 | -0.0751 |
| 3. Sense of Belonging | -0.0299 | -0.1011 | -0.0328 |
| d. Psychological Well-Being | -0.0410 | -0.0755 | -0.0333 |
| 1. Fullness vs. Emptiness of Life | -0.0463[c] | -0.0474 | -0.0684 |
| 2. Social Respect | -0.0096 | -0.1143[b] | 0.0415 |
| 3. Personal Freedom | -0.0448 | -0.0594 | 0.0326 |
| 4. Companionship | 0.0205 | -0.0118 | -0.0596 |
| 5. Tranquility vs. Anxiety | -0.0269 | -0.1072 | -0.1030 |
| 6. Self Approving vs. Guilty | -0.0347 | -0.0362 | -0.0464 |
| 7. Self Confidence | 0.0237 | -0.0582 | 0.1348[b] |
| 8. Elation vs. Depression | -0.0195 | -0.0077 | -0.0996 |

a. two-tailed t-test yields $p>0.100$ unless otherwise noted
b. t-test yields $p<0.100$
c. t-test yields $p<0.050$
d. t-test yields $p<0.010$
e. t-test yields $p<0.005$

TABLE 4.27 RELATIONSHIPS BETWEEN MAN'S TRAVEL TIME TO WORK AND INDEXES OF SATISFACTION: U.S.A., ISRAEL AND THE NETHERLANDS

| Satisfaction Index | Partial Correlation Coefficient[a] | | |
|---|---|---|---|
| | U.S.A. (n=825) | Israel (n=295) | Netherlands (n=215) |
| a. Housing Environment | -0.0766[c] | -0.1056 | -0.0833 |
| 1. Privacy | -0.0388 | 0.0276 | -0.0030 |
| 2. House & Lot | -0.0221 | 0.0193 | -0.0253 |
| 3. Household Responsibilities | -0.0932[d] | -0.1135[b] | -0.0086 |
| 4. Appearance | -0.0508 | -0.0958 | -0.0656 |
| b. Community Services | -0.0691[c] | -0.2374[c] | -0.0746 |
| 1. Parks | -0.0389 | -0.1491[c] | -0.1612[c] |
| 2. Schools | -0.0661 | -0.0939 | 0.1284 |
| 3. Security | -0.0265 | -0.0904 | -0.0056 |
| 4. Child Care | 0.0026 | 0.0297 | -0.0866 |
| 5. Transportation | -0.0790[c] | -0.0328 | -0.1463[b] |
| 6. Entertainment | -0.0612[b] | -0.0603 | -0.0222 |
| c. Social Patterns | -0.0614[b] | -0.0617 | 0.0536 |
| 1. Friendships | -0.0197 | -0.0739 | -0.0125 |
| 2. Group Activities | -0.0441 | 0.0422 | -0.0057 |
| 3. Sense of Belonging | -0.0602[b] | 0.0617 | 0.0784 |
| d. Psychological Well-Being | -0.0277 | -0.1565[c] | 0.0291 |
| 1. Fullness vs. Emptiness of Life | -0.0572[b] | 0.0543 | -0.0216 |
| 2. Social Respect | 0.0216 | 0.0728 | 0.0195 |
| 3. Personal Freedom | 0.0310 | -0.1383[c] | -0.0046 |
| 4. Companionship | -0.0631[b] | -0.1029 | 0.0948 |
| 5. Tranquility vs. Anxiety | 0.0385 | -0.1617[c] | -0.0313 |
| 6. Self Approving vs. Guilty | -0.0294 | -0.0886 | 0.0155 |
| 7. Self Confidence | 0.0432 | -0.0761 | 0.0155 |
| 8. Elation vs. Depression | -0.0275 | -0.1065 | 0.0029 |

a. two-tailed t-test yields $p>0.100$ unless otherwise noted
b. t-test yields $p<0.100$
c. t-test yields $p<0.050$
d. t-test yields $p<0.010$
e. t-test yields $p<0.005$

with the overall satisfaction of the housing environment ($p<0.050$) based primarily on the dissatisfaction indicated for household responsibilities ($p<0.100$).

The travel time to work of Israeli men also strongly supported expectations as it related negatively with overall satisfaction of community services, and psychological well-being (both $p<0.050$). These relationships were primarily based on negative scores for parks, personal freedom, and tranquility (all $p<0.050$). As in the United States, the Israeli men's travel time to work was negatively related to household responsibilities.

The Dutch data further supported our expectations, but with less statistical force. The dissatisfactions were centered around the negative scores for community services, notably parks ($p<0.050$) and transportation ($p<0.100$).

Discussion: The fact that in all three countries men's travel time to work has a comparable, if not greater, overall negative impact on women's satisfaction than women's travel time may reflect certain female-male dependencies based on traditional household role structures. For example, the negative relationship of men's travel time and women's satisfaction of the housing environment is based largely on dissatisfaction with household responsibilities - activities which involve home maintenance work, which traditionally has been considered men's work, and which may not be completed because of men's excessive time away from home while commuting to and from work. The same process may be functioning with the women's enjoyment of community services being dependent on men's availability to serve as an escort for those services such as parks and entertainment and culture. Thus, the effects of commutation patterns seem to reflect women's dependence on the presence of men for satisfaction of housing and community services in these three societies.

5. Population Size of Political Unit

Expectation: The population size of a political unit will vary inversely with satisfaction on all community service variables.

Findings: This expectation was strongly supported by our data from the United States and Israel and moderately supported by that from the Netherlands.

A comparison of four central city neighborhoods in San Jose (1975 population, 551,400) with four suburban neighborhoods in Los Gatos (1975 population, 23,900) and Cupertino (1975 population, 22,000) shows a significant difference in scores of overall satisfaction with community services at the 0.005 level favoring the smaller suburban communities (see Table 4.28). Indeed, the suburban communities had significantly higher scores for all specific indexes of satisfaction for group activities ($p<0.005$). Even when controlling for age of housing, it was found that the suburban communities consistently have higher scores of overall satisfaction with community services ($p<0.050$) and to a lesser extent with social patterns.[19] The central city excelled only with respect to the overall satisfaction with the housing environment in that housing in the older neighborhood attracted strong support for satisfaction with its appearance.

Data from Israeli suburban communities of Raanana and Kefar Sava (1978 populations of 24,700 and 35,300 respectively) near Tel-Aviv, when compared with that of Bavly, a neighborhood in Tel-Aviv (1978 population, 343,300) reveal an Israeli pattern similar to that of the United States. As Table 4.29 indicates, despite some disadvantages with regard to housing privacy and accessibility to entertainment, the suburban communities were found to be considerably more desirable than the central city neighborhood with respect to the community services of schools, security, child care and transportation ($p<0.050$), and especially with respect to overall index of psychological well-being ($p<0.001$). The latter result is based on a positive relationship between the suburban communities and almost all specific indexes of psychological well-being.

In the Netherlands, a comparison of three central city neighborhoods from The Hague (1979 population, 458,000) with the suburban community of Rijswijk (1980 population, 52,600) shows the suburban area to have a significant advantage in the community services specific indexes of parks ($p<0.100$) and entertainment ($p<0.005$) (see Table 4.30). This superiority is sufficient to influence a higher overall satisfaction with community services for the suburban area over the central city neighborhoods ($p<0.050$). However, unlike the United States, where housing satisfaction tended to be higher in an older neighborhood, such is not the

TABLE 4.28 MEAN SCORES OF INDEXES OF SATISFACTION: CENTRAL CITY AND SUBURBS: THE SAN JOSE AREA

| Satisfaction Index | Satisfaction Scores[a] | |
|---|---|---|
| | Central City (n=403) | Suburbs (n=422) |
| a. Housing Environment | 5.62 | 5.56 |
| 1. Privacy | 5.87 | 5.77 |
| 2. House & Lot | 5.61 | 5.52 |
| 3. Household Responsibilities | 5.21 | 5.22 |
| 4. Appearance | 5.81 | 5.72 |
| b. Community Services | 4.88 | 5.28[e] |
| 1. Parks | 4.69 | 5.23[e] |
| 2. Schools | 4.95 | 5.38[e] |
| 3. Security | 4.98 | 5.48[e] |
| 4. Child Care | 4.30 | 4.09[c] |
| 5. Transportation | 5.25 | 5.57[e] |
| 6. Entertainment | 4.99 | 5.95[e] |
| c. Social Patterns | 5.27 | 5.37 |
| 1. Friendships | 5.60 | 5.71 |
| 2. Group Activities | 5.11 | 5.39[e] |
| 3. Sense of Belonging | 5.07 | 5.01 |
| d. Psychological Well-Being | 6.94 | 6.94 |
| 1. Fullness vs. Emptiness of Life | 7.05 | 7.12 |
| 2. Social Respect | 7.45 | 7.51 |
| 3. Personal Freedom | 6.77 | 6.73 |
| 4. Companionship | 7.59 | 7.47 |
| 5. Tranquilty vs. Anxiety | 6.29 | 6.34 |
| 6. Self Approving vs. Guilty | 6.74 | 6.70 |
| 7. Self Confidence | 7.10 | 7.08 |
| 8. Elation vs. Depression | 6.66 | 6.65 |

a. two-tailed t-test for all score differences yields $p>0.100$ unless otherwise noted
b. t-test yields $p<0.100$
c. t-test yields $p<0.050$
d. t-test yields $p<0.010$
e. t-test yields $p<0.005$

TABLE 4.29 MEAN SCORES OF INDEXES OF SATISFACTION: CENTRAL CITY AND SUBURBS: THE TEL-AVIV AREA

| Satisfaction Index | Satisfaction Scores[a] | |
|---|---|---|
| | Central City (n=50) | Suburbs (n=115) |
| a. Housing Environment | 5.38 | 5.08 |
| 1. Privacy | 6.06[c] | 5.62 |
| 2. House & Lot | 5.70 | 5.50 |
| 3. Household Responsibilities | 4.88 | 4.49 |
| 4. Appearance | 5.08 | 4.68 |
| b. Community Services | 4.13 | 4.49 |
| 1. Parks | 4.68 | 4.76 |
| 2. Schools | 4.13 | 5.28[e] |
| 3. Security | 4.91 | 5.45[c] |
| 4. Child Care | 3.69 | 4.29[c] |
| 5. Transportation | 3.37 | 4.34[c] |
| 6. Entertainment | 4.88[e] | 2.79 |
| c. Social Patterns | 4.24 | 4.57 |
| 1. Friendships | 5.13 | 4.96 |
| 2. Group Activities | 3.16 | 3.49 |
| 3. Sense of Belonging | 4.55 | 5.34[b] |
| d. Psychological Well-Being | 6.37 | 7.15[e] |
| 1. Fullness vs. Emptiness of Life | 5.80 | 6.74[c] |
| 2. Social Respect | 7.50 | 7.81 |
| 3. Personal Freedom | 6.45 | 7.15[e] |
| 4. Companionship | 7.00 | 7.81[c] |
| 5. Tranquilty vs. Anxiety | 5.90 | 6.84[d] |
| 6. Self Approving vs. Guilty | 6.55 | 7.21[c] |
| 7. Self Confidence | 5.90 | 6.86[e] |
| 8. Elation vs. Depression | 5.85 | 6.72[e] |

a. two-tailed t-test for all score differences yields $p>0.100$ unless otherwise noted
b. t-test yields $p<0.100$
c. t-test yields $p<0.050$
d. t-test yields $p<0.010$
e. t-test yields $p<0.005$

TABLE 4.30 MEAN SCORES OF INDEXES OF SATISFACTION: CENTRAL CITY AND SUBURB: THE HAGUE AREA

| Satisfaction Index | Satisfaction Scores[a] | |
|---|---|---|
| | Central City (n=162) | Suburb. (n=53) |
| a. Housing Environment | 5.05 | 5.31[b] |
| 1. Privacy | 4.92 | 4.88 |
| 2. House & Lot | 5.33 | 5.94[d] |
| 3. Household Responsibilities | 5.01[c] | 4.46 |
| 4. Appearance | 4.92 | 5.96[e] |
| b. Community Services | 5.08 | 5.32[c] |
| 1. Parks | 5.24 | 5.65[b] |
| 2. Schools | 5.82 | 5.62 |
| 3. Security | 4.28 | 4.44 |
| 4. Child Care | 4.24 | 4.31 |
| 5. Transportation | 6.04 | 5.88 |
| 6. Entertainment | 4.81 | 5.71[e] |
| c. Social Patterns | 4.72 | 4.66 |
| 1. Friendships | 5.15 | 5.56 |
| 2. Group Activities | 4.76[b] | 4.24 |
| 3. Sense of Belonging | 4.16 | 4.13 |
| d. Psychological Well-Being | 6.57 | 6.41 |
| 1. Fullness vs. Emptiness of Life | 6.46 | 6.33 |
| 2. Social Respect | 6.81 | 6.65 |
| 3. Personal Freedom | 6.97 | 6.85 |
| 4. Companionship | 7.09 | 6.87 |
| 5. Tranquilty vs. Anxiety | 6.13 | 5.96 |
| 6. Self Approving vs. Guilty | 6.20 | 6.19 |
| 7. Self Confidence | 6.30 | 5.92 |
| 8. Elation vs. Depression | 6.64 | 6.50 |

a. two-tailed t-test for all score differences yields $p>0.100$ unless otherwise noted
b. t-test yields $p<0.100$
c. t-test yields $p<0.050$
d. t-test yields $p<0.010$
e. t-test yields $p<0.005$

case in the Netherlands. Old Leyenburg, the pre-1945 area, enjoys no statistical advantage over suburban Rijswijk.[20] It is in the newer, central city residential areas where satisfaction with the specific index of household responsibilities is higher ($p<0.050$) than in suburban Rijswijk, where housing costs tend to be higher.

However, with respect to other indexes of satisfaction with the housing environment, Rijswijk prevails over the central city neighborhoods in the overall satisfaction with housing environment index ($p<0.100$) as well as on the specific indexes of satisfaction with house and lot ($p<0.010$) and appearance ($p<0.005$); this pattern is very nearly replicated when Rijswijk is compared with Old Leyenburg.[21] No decisive pattern of superiority emerges for the suburban area when the satisfaction with social patterns or psychological well-being is considered.

Discussion: Prior to the passing of California's property tax limiting Proposition 13 in 1978, residents of political units with smaller and more homogeneous populations have generally registered greater satisfaction with community services than residents of larger political units.[22] This is a function of public service preferences enacted at the local level, moderated by the municipal financial ability to support such services. Although the financial resources for supporting suburban public services would not pose a problem for the middle-income households in this study, it would be a formidable obstacle to many central city households with modest incomes. The post-Proposition 13 environment in California has resulted in scarce resources for all local governments and the superiority enjoyed by suburban units of government in California would certainly be affected by the fiscal realities of the 1980s.

Our American findings suggest also that the superiority of smaller units for community service satisfaction may be related to the more participatory nature of local politics. This involvement in turn is likely to provide residents of small communities with a relatively greater control over and satisfaction with their public services. Perhaps, as indicated in other studies, this activity is a means of preserving strong community sovereignty within the metropolitan area.[23]

While many local public services are provided

by the national government in Israel, local residents do have a say about the character of such services and indeed about the planning of their entire communities.[24] As did their American counterparts, Israeli residents of smaller suburban political units registered higher satisfaction with the quality of public services such as schools. Moreover, the Israeli suburban satisfaction went beyond that of the United States to include psychological rewards. This seems to represent the underlying desire of many middle class families to trade off central city accessibility for the social, political, and psychological rewards of living in a relatively small suburban community.

In the Netherlands, local public services are funded by the central government, but administered locally, and one might expect a more uniform provision of local community services than in the United States. Such appears to be the case, for only satisfaction with entertaiment is uniformly superior in suburban Rijswijk. Not surprisingly, this is the least dependent on public resources of all community services; it is the chief determinant of the overall satisfaction with community services. Only when Rijswijk is compared with New Leyenburg does satisfaction with parks show a statistical superiority in favor of the suburb. This may be explained by New Leyenburg's greater distance from a large municipal park in The Hague; Old Leyenburg is marginally closer, while Bouwlust provides ample open spaces as part of its design as a highly planned community.

What role does housing choice play in this discussion? In the Netherlands, the answer is very little. Due to the continuing post-1945 housing shortage, 20% of households indicated their choice of housing was based on the fact that it was the only unit available at the time of their move. In contrast, California respondents were divided into two groups: those from the central city who focussed their decision on housing characteristics and those from the suburbs who stressed that public services, especially schools, were the chief determinant upon which their purchase decision was based.

It remains to be seen whether the provision of public services, both before and after Proposition 13, remains a significant factor in American housing choice. Prior to Proposition 13, it appears that all households were satisfied with the level of services which they had expected or that, over time, environmentally influences, such as housing considerations, took precedence over more distant neighborhood qualities.[25] This supports Michelson's concept that residential satisfaction may reflect not only a mix of self-selection and the behavioral patterns emerging from environmental situations, but also the degree of congruence between a current situation and long-range environmental aspirations. The role of community services as a factor in American housing choice subsequent to Proposition 13 remains to be seen.

## 6. Design and Site Plan Characteristics

Expectation: The greater the degree of planned characteristics in the site plan of a housing development, the greater the level of satisfaction on all dependent variables.

Findings: Data collected in the United States is partially supportive of expectations, while that from Israel is contradictory and that from the Netherlands is ambiguous. Comparing planned areas in the San Jose region (comprising planned unit developments of single-family homes and planned multi-family condominiums) to less planned areas (made up of conventional, single-family suburban neighborhoods), Table 4.31 indicates a significantly higher overall satisfaction with the housing environment ($p<0.010$) and community services ($p<0.050$) for the planned areas, and greater overall satisfaction with social patterns for the less-planned neighborhoods ($p<0.050$). Inconclusive results were manifested for measures of psychological well-being. The desirability of planned areas was based on the specific satisfactions with household responsibilities for the housing environment and sense of security for community services, both significant at the 0.005 level; while the unexpected superiority of the less-planned areas centered around specific satisfactions of friendships ($p<0.010$) and sense of belonging ($p<0.005$) for social patterns. The inconclusive finding concerning the influence of planned environments on psychological well-being results from two compensating relationships: the higher satisfaction of personal freedom for residents of planned areas ($p<0.005$); and greater fullness of life for less-planned area residents ($p<0.010$).

However, when Israeli data are examined,

TABLE 4.31 MEAN SCORES OF INDEXES OF SATISFACTION: PLANNED AND LESS PLANNED AREAS: THE SAN JOSE AREA

| Satisfaction Index | Satisfaction Scores[a] | |
|---|---|---|
| | Planned Areas (n=406) | Less Planned Areas (n=419) |
| a. Housing Environment | 5.67[d] | 5.51 |
| 1. Privacy | 5.78 | 5.85 |
| 2. House & Lot | 5.51 | 5.62 |
| 3. Household Responsibilities | 5.63[e] | 4.81 |
| 4. Appearance | 5.77 | 5.76 |
| b. Community Services | 5.16[c] | 5.04 |
| 1. Parks | 4.98 | 4.96 |
| 2. Schools | 5.21 | 5.14 |
| 3. Security | 5.45[e] | 5.04 |
| 4. Child Care | 4.24 | 4.12 |
| 5. Transportation | 5.42 | 5.41 |
| 6. Entertainment | 5.52 | 5.45 |
| c. Social Patterns | 5.23 | 5.41[c] |
| 1. Friendships | 5.55 | 5.76[d] |
| 2. Group Activities | 5.32 | 5.19 |
| 3. Sense of Belonging | 4.82 | 5.25[e] |
| d. Psychological Well-Being | 6.94 | 6.93 |
| 1. Fullness vs. Emptiness of Life | 6.98 | 7.18[c] |
| 2. Social Respect | 7.45 | 7.51 |
| 3. Personal Freedom | 6.91[e] | 6.59 |
| 4. Companionship | 7.54 | 7.52 |
| 5. Tranquilty vs. Anxiety | 6.32 | 6.21 |
| 6. Self Approving vs. Guilty | 6.68 | 6.77 |
| 7. Self Confidence | 7.10 | 7.08 |
| 8. Elation vs. Depression | 6.64 | 6.66 |

a. two-tailed t-test for all score differences yields $p>0.100$ unless otherwise noted
b. t-test yields $p<0.100$
c. t-test yields $p<0.050$
d. t-test yields $p<0.010$
e. t-test yields $p<0.005$

TABLE 4.32 MEAN SCORES OF INDEXES OF SATISFACTION: PLANNED AND LESS PLANNED AREAS: ISRAEL

| Satisfaction Index | Satisfaction Scores[a.] | |
|---|---|---|
| | Planned Areas (n=180) | Less Planned Areas (n=115) |
| a. Housing Environment | 4.55 | 5.43[e.] |
| 1. Privacy | 5.13 | 6.00[e.] |
| 2. House & Lot | 4.73 | 5.94[e.] |
| 3. Household Responsibilities | 3.98 | 4.72[e.] |
| 4. Appearance | 4.37 | 5.05[e.] |
| b. Community Services | 3.78 | 4.69[e.] |
| 1. Parks | 4.28 | 4.95[e.] |
| 2. Schools | 4.71 | 5.07[b.] |
| 3. Security | 4.98 | 5.49[e.] |
| 4. Child Care | 3.69 | 3.99 |
| 5. Transportation | 3.36 | 4.56[e.] |
| 6. Entertainment | 2.60 | 3.65[e.] |
| c. Social Patterns | 4.32 | 4.67[e.] |
| 1. Friendships | 4.54 | 5.34[e.] |
| 2. Group Activities | 3.49 | 3.57 |
| 3. Sense of Belonging | 4.99 | 5.23 |
| d. Psychological Well-Being | 6.87 | 7.00 |
| 1. Fullness vs. Emptiness of Life | 6.63 | 6.48 |
| 2. Social Respect | 7.82 | 7.79 |
| 3. Personal Freedom | 6.61 | 7.11 |
| 4. Companionship | 7.38 | 7.71 |
| 5. Tranquilty vs. Anxiety | 6.48 | 6.87 |
| 6. Self Approving vs. Guilty | 7.06 | 7.04 |
| 7. Self Confidence | 6.52 | 6.60 |
| 8. Elation vs. Depression | 6.52 | 6.64 |

a. two-tailed t-test for all score differences yields $p>0.100$ unless otherwise noted
b. t-test yields $p<0.100$
c. t-test yields $p<0.050$
d. t-test yields $p<0.010$
e. t-test yields $p<0.005$

TABLE 4.33 MEAN SCORES OF INDEXES OF SATISFACTION: PLANNED AND LESS PLANNED AREAS: THE NETHERLANDS

| Satisfaction Index | Satisfaction Scores[a] | |
|---|---|---|
| | Planned Areas (n=113) | Less Planned Areas (n=102) |
| a. Housing Environment | 5.08 | 5.17 |
| 1. Privacy | 5.05[c] | 4.78 |
| 2. House & Lot | 5.40 | 5.58 |
| 3. Household Responsibilities | 5.10[e] | 4.62 |
| 4. Appearance | 4.77 | 5.60[e] |
| b. Community Services | 5.03 | 5.25[c] |
| 1. Parks | 5.20 | 5.49[b] |
| 2. Schools | 5.86 | 5.70 |
| 3. Security | 4.27 | 4.37 |
| 4. Child Care | 4.06 | 4.45[b] |
| 5. Transportation | 6.16[c] | 5.84 |
| 6. Entertainment | 4.70 | 5.38[e] |
| c. Social Patterns | 4.78 | 4.64 |
| 1. Friendships | 5.26 | 5.25 |
| 2. Group Activities | 4.96[e] | 4.30 |
| 3. Sense of Belonging | 4.08 | 4.23 |
| d. Psychological Well-Being | 6.60 | 6.47 |
| 1. Fullness vs. Emptiness of Life | 6.44 | 6.42 |
| 2. Social Respect | 6.85 | 6.69 |
| 3. Personal Freedom | 6.97 | 6.89 |
| 4. Companionship | 7.13 | 6.91 |
| 5. Tranquilty vs. Anxiety | 6.18 | 5.99 |
| 6. Self Approving vs. Guilty | 6.12 | 6.28 |
| 7. Self Confidence | 6.34 | 6.08 |
| 8. Elation vs. Depression | 6.75[b] | 6.47 |

a. two-tailed t-test for all score differences yields $p>0.100$ unless otherwise noted
b. t-test yields $p<0.100$
c. t-test yields $p<0.050$
d. t-test yields $p<0.010$
e. t-test yields $p<0.005$

less-planned areas enjoy a convincing superiority over planned areas (see Table 4.32). Excluding psychological well-being, where no differences could be discerned, less-planned areas were strongly preferred by respondents on the three other overall indexes of satisfaction: housing environment, community services and social patterns (all $p<0.010$). The greatest superiority of the less-planned areas was in the quality of housing where all specific indexes of satisfaction were significant at the 0.005 level.

Dutch planned areas (see Table 4.33) have provided housing favored by respondents for superior privacy ($p<0.050$) as well as ease of managing household responsibilities ($p<0.005$), whereas less-planned areas provide greater satisfaction with the appearance of the dwelling unit ($p<0.005$). However, in the area of community services, less planned areas are superior in overall satisfaction ($p<0.050$), satisfaction with parks and child care ($p<0.100$ for both) and satisfaction with entertainment ($p<0.005$). Yet, planned areas do provide greater satisfaction with transportation ($p<0.050$). This reversal of the American pattern continues with greater satisfaction with the specific index of group activities ($p<0.005$) demonstrated in planned areas. No conclusive data is provided by the various indexes of psychological well-being.

If we attempt to control for age of housing, the basic relationship between planned and less-planned housing remains essentially unchanged for the United States and Israel, but not for Holland.[27] In the Netherlands planned housing environments 7 to 17 years old (i.e., built from 1961-1971) have significantly higher scores on overall satisfaction with psychological well-being ($p<0.050$) as well as on three specific psychological indexes ($p<0.050$) than did housing built prior to 1960.

If adjacent locations are considered and controlled for age of dwelling units and density, it is found that in the San Jose area the planned area continues to enjoy a substantial advantage in terms of overall satisfaction with community services ($p<0.005$).[28] Indeed, the planned areas excel in five of the six specific indexes in this category. A comparison of adjacent areas in Israel and the Netherlands essentially replicates the significantly greater resident satisfaction with less-planned areas.[29]

<u>Discussion</u>: The concept of planned characteristics for residential development covers a broad range of possibilities. In the United States, so-called 'planned communities' attempt to package housing, recreational and other services in order to compete successfully within a given housing submarket. Therefore, it appears that planned environments have provided more efficient and rewarding housing environments than less-planned areas. However, depending on factors such as design and the social characteristics of residents, planned areas may be less socially rewarding than conventional single-family areas. An explanation for the relatively low degree of social interaction in planned areas is that a higher percentage of respondents were employed outside the household (64% in planned areas and 44% in less-planned neighborhoods). It seems likely that respondents in the planned environments simply have less time and energy for socializing. Additionally, the superior mix of and accessibility to community services provided by planned environments fulfill a need in the housing market that less-planned communities cannot offer. The greater level of satisfaction with household responsibilities ($p<0.005$) in planned areas also reflects the collective maintenance of outside areas which is characteristic of the planned environments in the San Jose area sample. With nearly two-thirds of respondents in planned areas involved in outside employment, the time and resources that would have to be expended on such activities increase the attractiveness of the planned environment.

The Israeli results, especially the overwhelming superiority of housing satisfaction in less-planned areas, reflect the fact that planned areas in Israel are usually comprised of public housing. Such housing, while adequate, is frugally designed and has limited interior amentities in terms of floor space, equipment and overall construction. For example, in the Tel-Aviv sample less-planned housing units had about 15 percent more floor space than planned units. Also many planned housing areas were developed on less expensive and more remote parcels with respect to community facilities than were market-oriented less-planned privately developed housing.[30] This remoteness of planned areas may help explain why respondents in such areas experience about twice the travel time to work (20 compared to 10 minutes) and lower satisfaction with friendships than do their counterparts in less-planned areas. Thus, in Israel it appears that the government planning process for housing may have been overwhelmed by the considerations of population pressure and cost constraints in land acquisition.

In the Netherlands, the superiority of planned environments takes on a different meaning from those in the United States. With more compact and heterogeneous metropolitan areas, the concept of a 'planned environment' in Holland covers a more narrow range of possibiities than in the United States. The superiority of a planned, inlying area (New Leyenburg) over an adjacent older one (Old Leyenburg) can be traced as much to the age and condition of the housing as to planned characteristics. Approximately 30 to 40 years separate the construction time of the two neighborhoods, and higher standards of living and superior accommodations characterize the newer housing. In the two outlying communities, the suburban location of Rijswijk provides for a more pleasant overall environment, although Bouwlust, the planned neighborhood in The Hague, does demonstrate greater satisfaction with household responsibilities than does suburban Rijswijk as well as greater satisfaction with the social patterns specific index of group activities.[31]

The greater satisfaction scores on these two items for planned Bouwlust (c. 1959) stem from the fact that, while all its units enjoy a government subsidy, not all of those in Rijswijk do, and indeed some respondents may not be receiving any government subsidy due to social status. This accounts for the noticeably higher level of satisfaction with the specific index of entertainment ($p<0.005$), a fact that was also influenced by the better access to the center of The Hague enjoyed by Rijswijk.

NOTES

[1] Based on data presented in Eli Borukhov, Yona Ginsberg and Elia Werczberger, 'Housing Prices and Housing Preferences in Israel', Urban Studies vol.15, no.2 (1978), pp.187-200; and Israeli Central Bureau of Statistics, Statistical Abstract of Israel: 1978 (Silvan Press, Jerusalm, 1978).

[2] Jan Van Weesep, Production and Allocation: The Case of the Netherlands. Geografische on Planologische Notities, no.11 (Free University of Amsterdam, Institute of Geography and Planning, Amsterdam, 1982), p.50.

[3] Nathan Glazer and Daniel P. Moynihan, Beyond the Melting Pot (MIT Press, Cambridge, Mass., 1963).

[4] Claude S. Fischer, To Dwell Among Friends: Personal Networks in Town and City (The University of Chicago Press, Chicago, 1982), pp.202-8.

[5] Ibid., pp.255-8.

[6] Angus Campbell, et al., The Quality of American Life: Perceptions, Evaluation, and Satisfactions (Russell Sage Foundation, New York, 1976).

[7] John F. Long, Population Decentralization in the United States (U.S. Bureau of the Census, Washington, D.C., 1981), Chapter 2.

[8] John J. Palen, 'The Urban Nexus: Toward the Year 2000', in Amos A. Hawley (ed.), Societal Growth (Free Press, New York, 1979), pp.141-56.

[9] Campbell, et al., The Quality of American Life, pp.411-4; and Helen Z. Lopata, Occupation Housewife (Oxford University Press, New York, 1971), p.71.

[10] Donald N. Rothblatt, Daniel J. Garr and Jo Sprague, The Suburban Environment and Women (Praeger, New York, 1979), Chapter 4.

[11] Ibid.

[12] Lynne M. Jones, 'The Labor Force Participation of Married Women', unpublished Master's Thesis, University of California, Berkeley, 1974.

[13] Claude S. Fischer, To Dwell Among Friends: Personal Networks in Town and City.

[14] Donald N. Rothblatt and Daniel J. Garr, Comparative Suburban Data (San Jose State University, San Jose, 1983).

[15] Amiram Gonen, 'The Suburban Mosaic in Israel', in D. H. K. Amiran and Y. Ben-Ariel (eds.), Geography in Israel, (The Israel National Committee, Jerusalem, 1976).

[16] Naomi Carmon and Bilha Mannheim, 'Housing as a Tool of Social Policy', Social Forces, vol. 58 no.2 (1979), pp.336-521.

[17] Norman Berdichevsky, 'The Persistence of the Yemeni Quarter in an Israeli Town', in Ernest Krausz, (ed.), Studies of Israeli Society: Volume 1 (Transaction Books, London, 1980), pp.73-97.

[18] Netherlands, Central Bureau of Statistics, Statistical Yearbook of the Netherlands: 1980 (Staatsuitgeverij, The Hague, 1981); Netherlands, Ministry of Housing and Physical Planning, Housing Production in the Netherlands (The Hague, 1978); Netherlands, Social and Cultural Planning Office, Social and Cultural Report: 1978 (The Hague, 1978).

[19] Rothblatt, Garr and Sprague, The Suburban Environment, Chapter 4.
[20] Rothblatt and Garr, Comparative Suburban Data, Chapter 2.
[21] Ibid.
[22] Donald C. Dahmann, 'Subjective Assessment of Neighborhood Quality by Size' Urban Studies, vol.20, no.1 (1983), pp.31-45.
[23] John R. Logan and Mark Schneider, 'Governmental Organization and City/Suburb Income Inequality, 1960-1970', Urban Affairs Quarterly, vol.17, no.3 (1982), pp.303-18.
[24] Nathaniel Lichfield, 'The Israeli Physical Planning System: Some Needed Changes', The Israel Annual of Public Administration, vol.15 (1976), pp.35-58.
[25] Rothblatt, Garr and Sprague, The Suburban Environment and Women, Chapter 5 .
[26] William Michelson, Man and His Urban Environment (Addison-Wesley, Reading, Mass., 1975).
[27] Rothblatt and Garr, Comparative Suburban Data, Chapter 3 .
[28] Rothblatt, Garr and Sprague, The Suburban Environment and Women, Chapter 5.
[29] Rothblatt and Garr, Comparative Suburban Data, Chapter 3.
[30] Borukhov, Ginsberg, and Werczberger, 'Housing Prices and Housing Preferences in Israel; Gonen, 'The Suburban Mosaic'.
[31] Rothblatt and Garr, Comparative Suburban Data, Chapter 4 .

# CHAPTER 5

# SUMMARY AND IMPLICATIONS FOR METROPOLITAN DEVELOPMENT IN THE INDUSTRIALIZED WORLD

This chapter presents a summary of study findings and their public policy implications for improving the quality of metropolitan living environments in the industrialized world.

## A. STUDY SUMMARY

In this section the comparative influence of clusters of independent variables and their impact on the global aspects of the quality of life indexes will be discussed. As noted in Chapter 3, these relationships were based primarily on partial correlation analysis and t-tests. Multiple regression analysis was not utilized for estimating the relative importance of the independent variables, due to its low explanatory power (<10%).

For the American study there seems to be a balance between environmental and social influences on overall housing environment satisfaction (see Figure 5.1). Table 5.1 reveals that all influence clusters had about a 70 to 80 percent success rate of significant independent variables at the 0.050 level, with the exception of subcultural influences. It appears that life cycle was the most important social influence and suburban environmental factors represented the single most telling set of independent variables. The variables with the greatest interaction with overall housing satisfaction are marital status, number of children and income.

Israeli results concerning overall housing environment satisfaction show a similar pattern to that of the United States. Figure 5.2 indicates that Israeli social factors, with the exception of subcultural influences, exert an impact on overall housing satisfaction similar to that of suburban

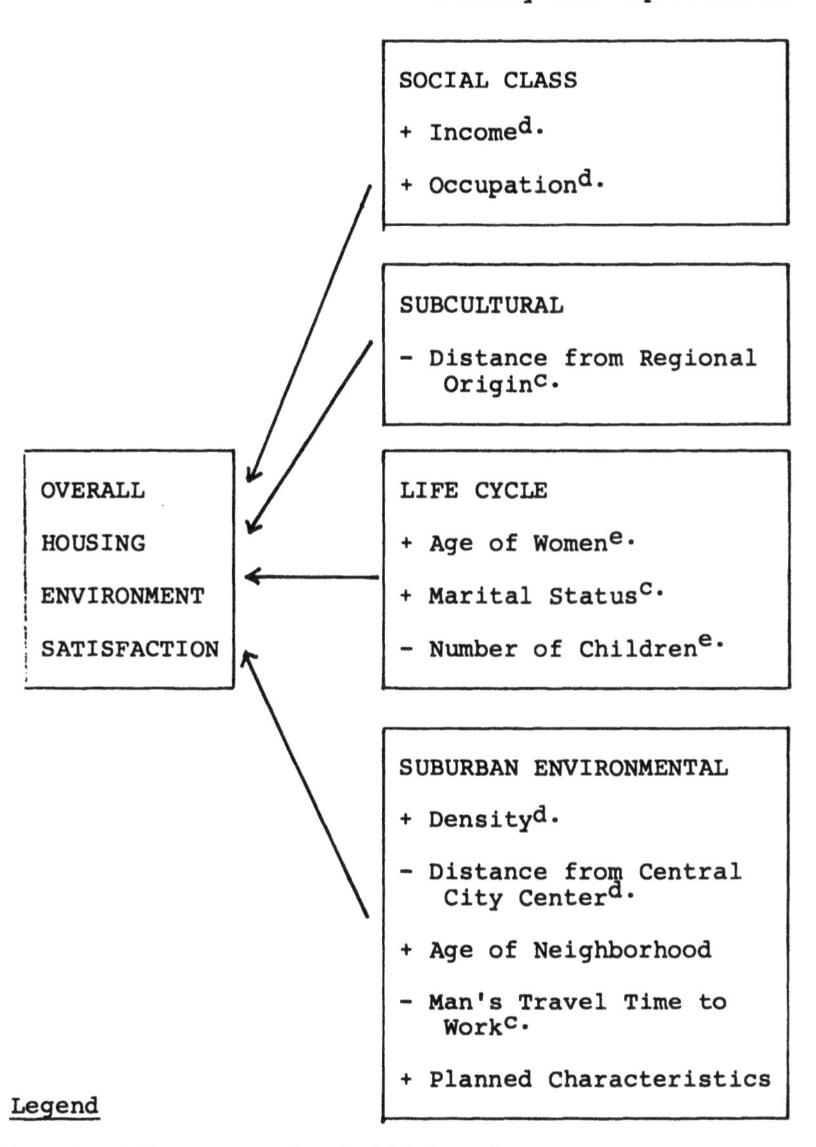


Legend

b. significant at the 0.100 level
c. significant at the 0.050 level
d. significant at the 0.010 level
e. significant at the 0.005 level

FIGURE 5.1 COMPARATIVE INFLUENCE OF INDEPENDENT VARIABLE CATEGORIES ON OVERALL HOUSING ENVIRONMENT SATISFACTION IN THE UNITED STATES

TABLE 5.1 COMPARATIVE SUCCESS RATE OF INFLUENCE OF INDEPENDENT VARIABLE CATEGORIES IN THE UNITED STATES

| Overall Index of Satisfaction | Percent Significant[a] Independent Variables in each Category | | | |
|---|---|---|---|---|
| | Social Class | Subcultural | Life Cycle | Suburban Environmental |
| Housing | 67 | 25 | 75 | 83 |
| Environment | 67[b] | 25[b] | 75[b] | 83[b] |
| Community | 33 | 25 | 25 | 83 |
| Services | 33[b] | 25[b] | 0[b] | 83[b] |
| Social | 33 | 25 | 25 | 67 |
| Patterns | 33[b] | 25[b] | 25[b] | 50[b] |
| Psychological | 33 | 25 | 75 | 17 |
| Well-Being | 0[b] | 25[b] | 75[b] | 0[b] |

a. Percentage independent variables found significant at the 0.100 level unless otherwise noted
b. Percentage independent variables found significant at the 0.050 level

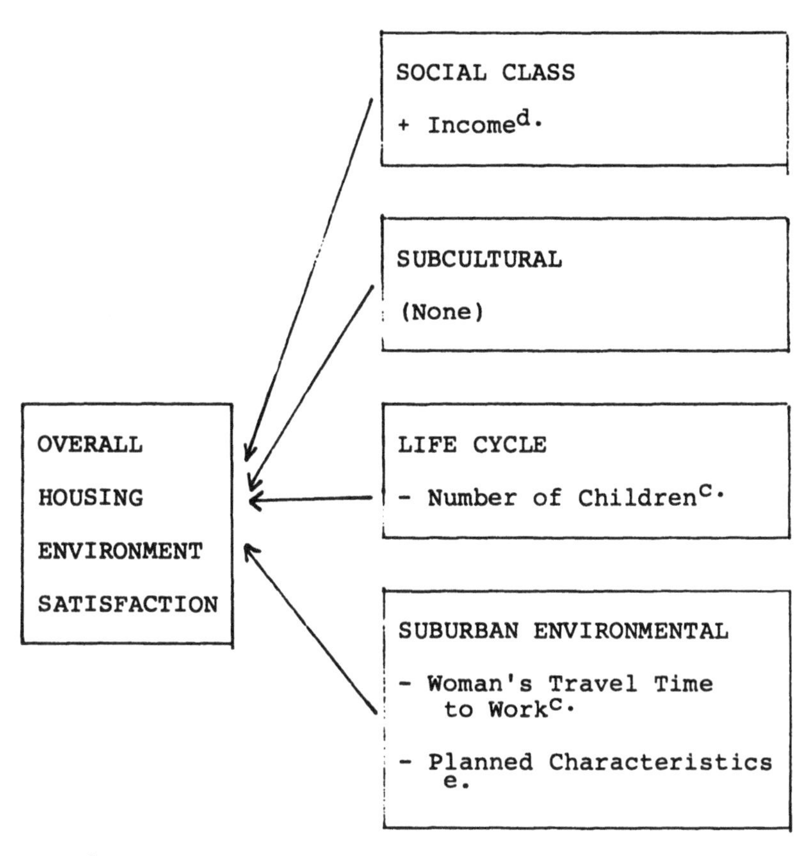


<u>Legend</u>

b. significant at the 0.100 level
c. significant at the 0.050 level
d. significant at the 0.010 level
e. significant at the 0.005 level

FIGURE 5.2 COMPARATIVE INFLUENCE OF INDEPENDENT VARIABLE CATEGORIES ON OVERALL HOUSING ENVIRONMENT SATISFACTION IN ISRAEL

TABLE 5.2 COMPARATIVE SUCCESS RATE OF INFLUENCE OF INDEPENDENT VARIABLE CATEGORIES IN ISRAEL

| Overall Index of Satisfaction | Percent Significant[a] Independent Variables in each Category | | | |
|---|---|---|---|---|
| | Social Class | Subcultural | Life Cycle | Suburban Environmental |
| Housing Environment | 33<br>33[b] | 0<br>0[b] | 25<br>25[b] | 33<br>33[b] |
| Community Services | 0<br>0[b] | 75<br>50[b] | 0<br>0[b] | 83<br>83[b] |
| Social Patterns | 67<br>67[b] | 25<br>25[b] | 50<br>50[b] | 33<br>17[b] |
| Psychological Well-Being | 67<br>67[b] | 25<br>25[b] | 50<br>50[b] | 83<br>67[b] |

a. Percentage independent variables found significant at the 0.100 level unless otherwise noted.
b. Percentage independent variables found significant at the 0.050 level

environmental influences. Table 5.2 shows that these influences were less than that in the United States as their success rates were only 33 percent at the 0.050 level. The single most influential variable is planned characteristics.

As Figure 5.3 reveals, Dutch findings are very pronounced in favor of the importance of suburban environmental influences on overall housing satisfaction. As Table 5.3 demonstrates, only environmental influences have a significant impact on housing satisfaction with a success rate of 75 percent at the 0.050 level. Here distance from the central city center is the most important variable.

When American overall community services satisfaction is examined (see Figure 5.4), it appears that the suburban environmental cluster becomes the overwhelmingly dominant set of influences. Indeed, if variables significant at the 0.050 level are scrutinized, educational level and degree of urbanness are the only social influences on community service satisfaction. Other important single variables with high correlations are the environmental variables concerned with community size and degree of planned characteristics.

Again, Israeli findings pertinent to community services satisfaction echo those from the United States. As Figure 5.5 shows, at the 0.050 level of significance, Israeli satisfaction with community services is entirely dependent on suburban environmental influences, with the exception of two subcultural variables, ethnic identity and length of national residence. The most influential environmental variables are women's travel time to work and degree of planned characteristics.

In a modest fashion, the Netherlands data reflects the dominance of environmental factors over social influences on community services satisfaction (see Figure 5.6). If analysis is limited to variables significant at the 0.050 level, only the environmental influences of community size and planned characteristics are of importance.

Figure 5.7 indicates a more equitable balance between social and environmental factors influencing overall social patterns satisfaction in the United States. Here the most important variables related to overall social satisfaction are concerned with education, marital status, and to some extent, time - length of residency, age of neighborhood, and women's travel time to work.

Unlike the American study, Israeli findings yielded a dominance of social over environmental

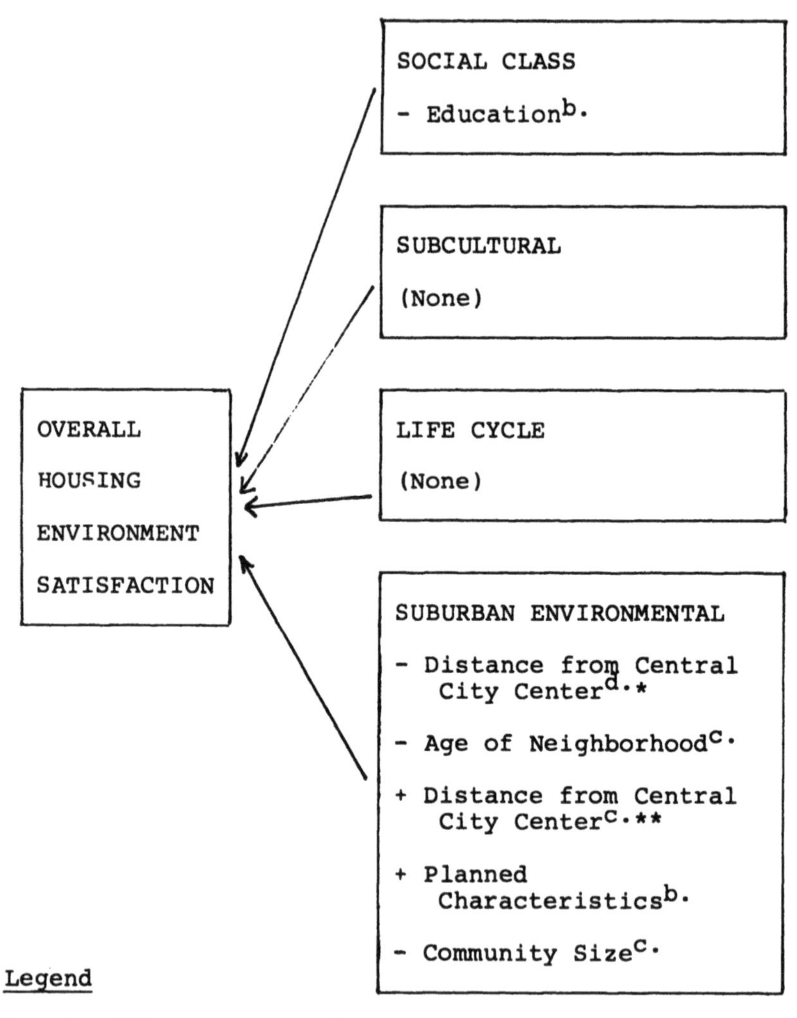


b. significant at the 0.100 level
c. significant at the 0.050 level
d. significant at the 0.010 level
e. significant at the 0.005 level

* Planned Neighborhoods only
** Less Planned Neighborhoods only

FIGURE 5.3 COMPARATIVE INFLUENCE OF INDEPENDENT VARIABLE CATEGORIES ON OVERALL HOUSING ENVIRONMENT SATISFACTION IN THE NETHERLANDS

TABLE 5.3 COMPARATIVE SUCCESS RATE OF INFLUENCE OF INDEPENDENT VARIABLE CATEGORIES IN THE NETHERLANDS

| Overall Index of Satisfaction | Percent Significant[a] Independent Variables in each Category | | | |
|---|---|---|---|---|
| | Social Class | Subcultural | Life Cycle | Suburban Environmental |
| Housing Environment | 33<br>0[b] | 0<br>0[b] | 0<br>0[b] | 83<br>67[b] |
| Community Services | 0<br>0[b] | 0<br>0[b] | 50<br>0[b] | 33<br>33[b] |
| Social Patterns | 33<br>0[b] | 0<br>0[b] | 0<br>0[b] | 17<br>0[b] |
| Psychological Well-Being | 0<br>0[b] | 0<br>0[b] | 0<br>0[b] | 33<br>17[b] |

a. Percentage independent variables found significant at the 0.100 level unless otherwise noted.
b. Percentage independent variables found significant at the 0.050 level

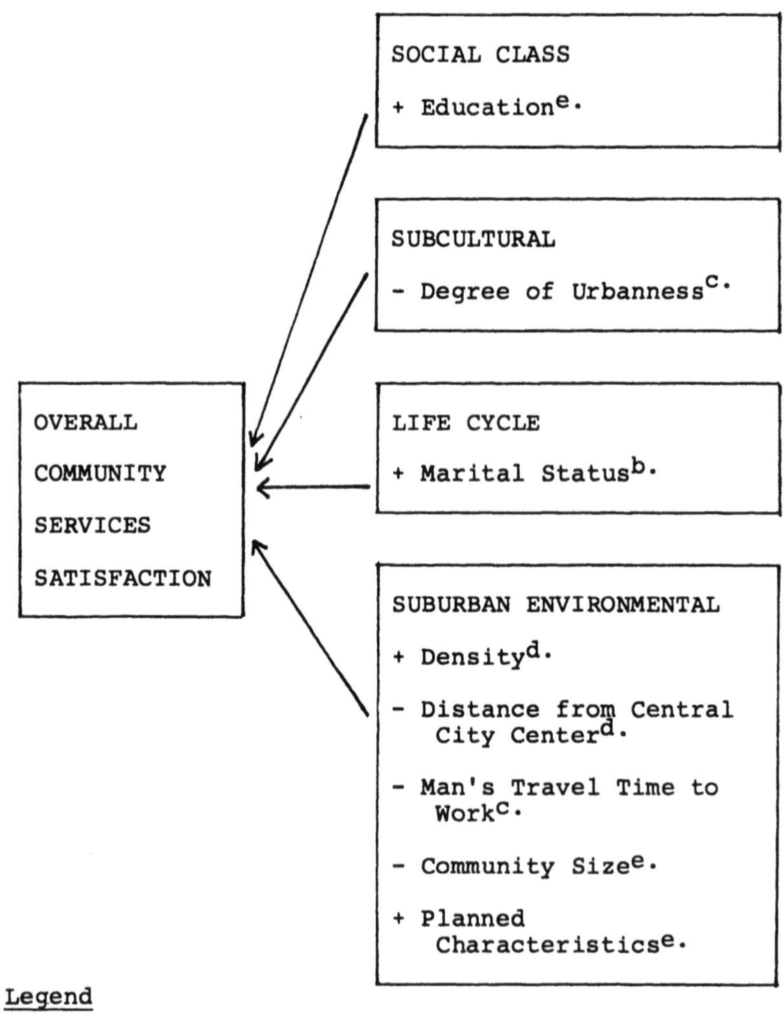


Legend

b. significant at the 0.100 level
c. significant at the 0.050 level
d. significant at the 0.010 level
e. significant at the 0.005 level

FIGURE 5.4 COMPARATIVE INFLUENCE OF INDEPENDENT VARIABLE CATEGORIES ON OVERALL COMMUNITY SERVICES SATISFACTION IN THE UNITED STATES

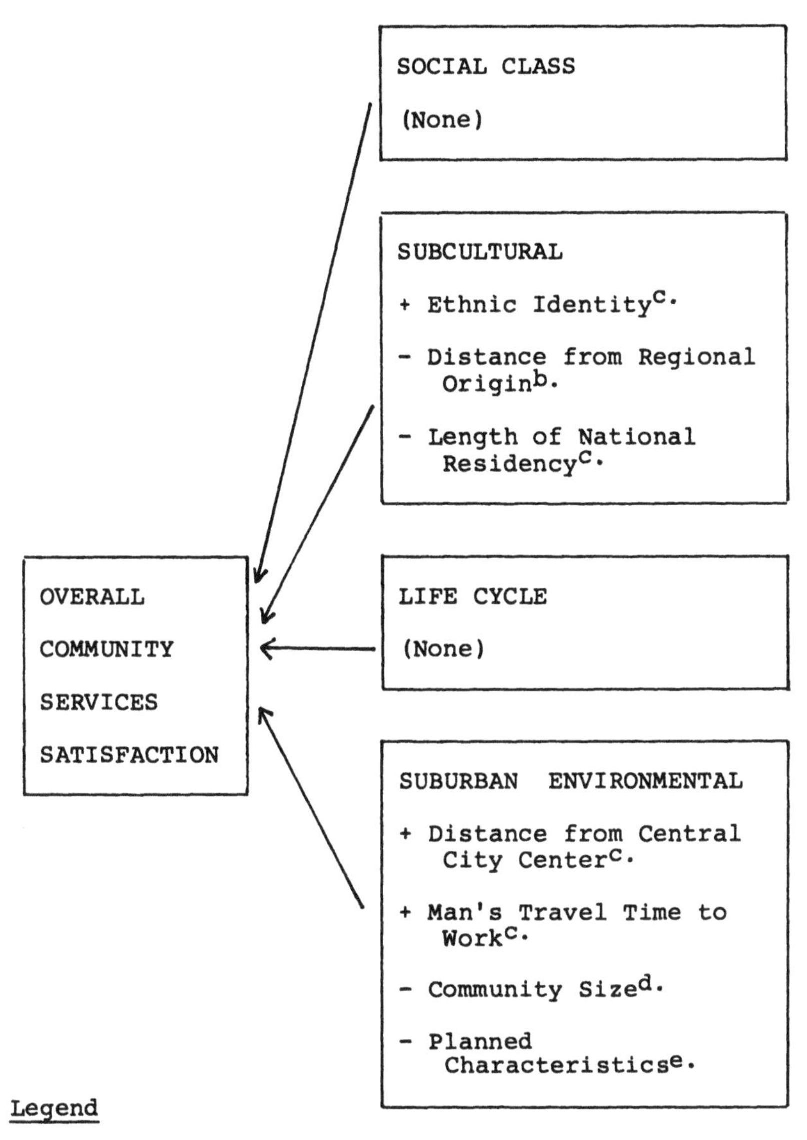


Legend

b. significant at the 0.100 level
c. significant at the 0.050 level
d. significant at the 0.010 level
e. significant at the 0.005 level

FIGURE 5.5 COMPARATIVE INFLUENCE OF INDEPENDENT VARIABLE CATEGORIES ON OVERALL COMMUNITY SERVICES SATISFACTION IN ISRAEL

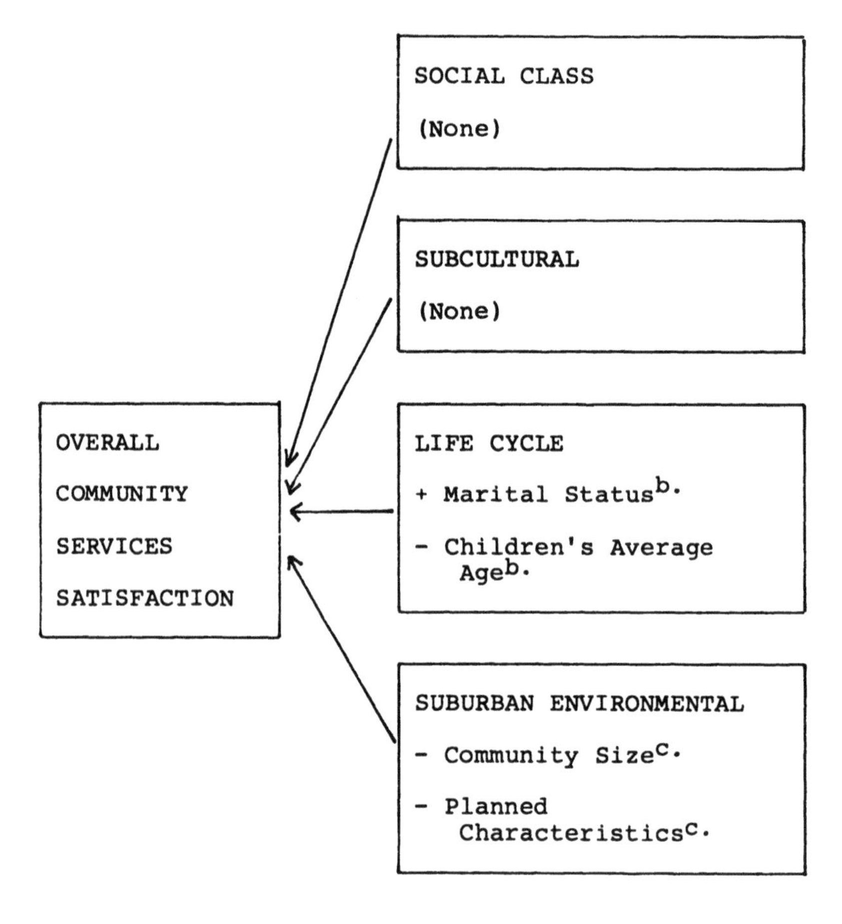


Legend

b. significant at the 0.100 level
c. significant at the 0.050 level
d. significant at the 0.010 level
e. significant at the 0.005 level

FIGURE 5.6 COMPARATIVE INFLUENCE OF INDEPENDENT VARIABLE CATEGORIES ON OVERALL COMMUNITY SERVICES SATISFACTION IN THE NETHERLANDS

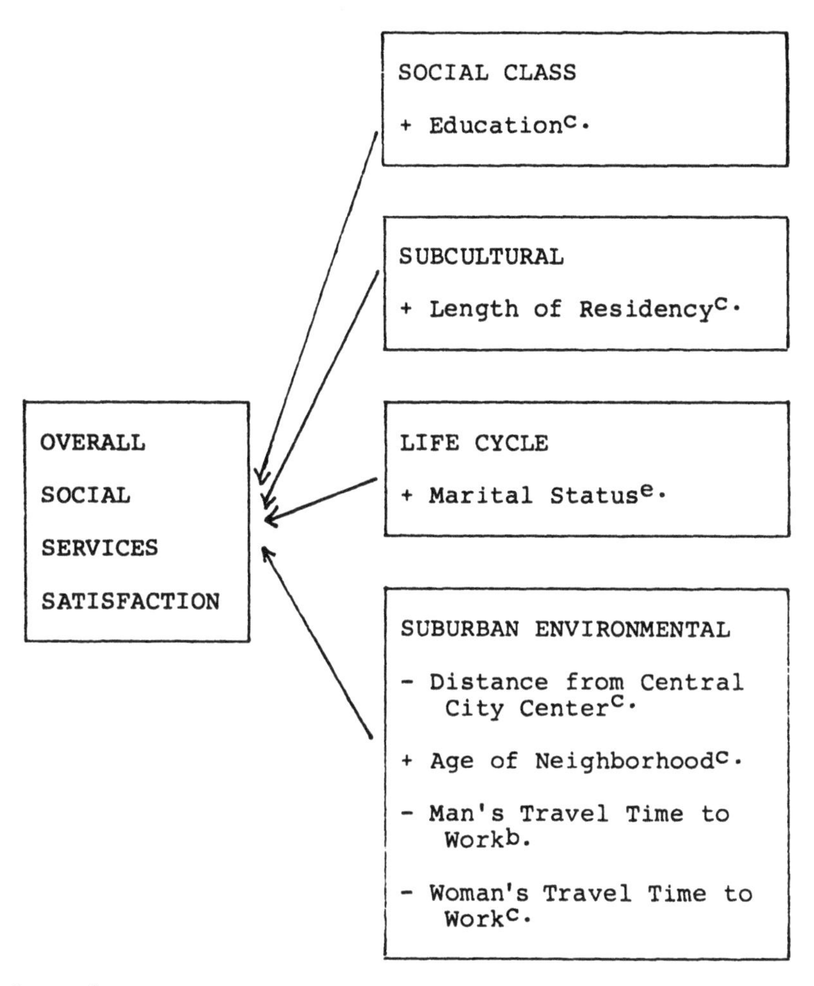


Legend

b. significant at the 0.100 level
c. significant at the 0.050 level
d. significant at the 0.010 level
e. significant at the 0.005 level

FIGURE 5.7 COMPARATIVE INFLUENCE OF INDEPENDENT VARIABLE CATEGORIES ON OVERALL SOCIAL PATTERNS SATISFACTION IN THE UNITED STATES

factors insofar as overall social patterns satisfaction is concerned. As shown on Table 5.2 and Figure 5.8, at the 0.050 level every social cluster of variables has a success rate substantially greater than that of environmental factors. The most prominent influences of social patterns satisfaction are income, women's age and number of children. In contrast, the Netherlands yielded an almost neutral influence for all variables on social patterns satisfaction. As Figure 5.9 reveals, only two variables, occupation, and distance from the central city center, had only a mild impact on social satisfaction at the 0.100 level.

As shown in Figure 5.10, in the United States, overall psychological well-being is almost entirely dominated by social influences, the most important of which is the life cycle cluster of variables. Only the modest 0.100 level relationship of age of neighborhood represents the sole environmental influence on psychological well-being. The individual variables which appear to have the greatest correlations with psychological well-being are ethnic identity and age of woman. Figure 5.11 indicates that the variables influencing psychological well-being in Israel differ from those in the United States in two basic ways: first, there is a rough parity of importance between social and environmental factors; and second, every cluster of variables had a substantial impact on psychological well-being (at least a 50 percent success rate at the 0.050 level for all clusters except subcultural, which had a 25 percent rate of success). In almost complete opposition to the American findings, the Dutch results show no social influence on psychological well-being. As Figure 5.12 demonstrates, only two environmental variables, distance from the central city center and planned characteristics, were significant at the 0.100 level.

In sum, it appears that in the United States environmental factors have the greatest impact on satisfaction with the housing environment and with community services; a moderate influence on that with social patterns; and a marginal effect on that with psychological well-being. In contrast, social factors are overwhelmingly the most significant influences on psychological well-being; moderately important for satisfaction with the housing environment and with social patterns; and more peripherally related to satisfaction with community services. The principal set of social forces appears

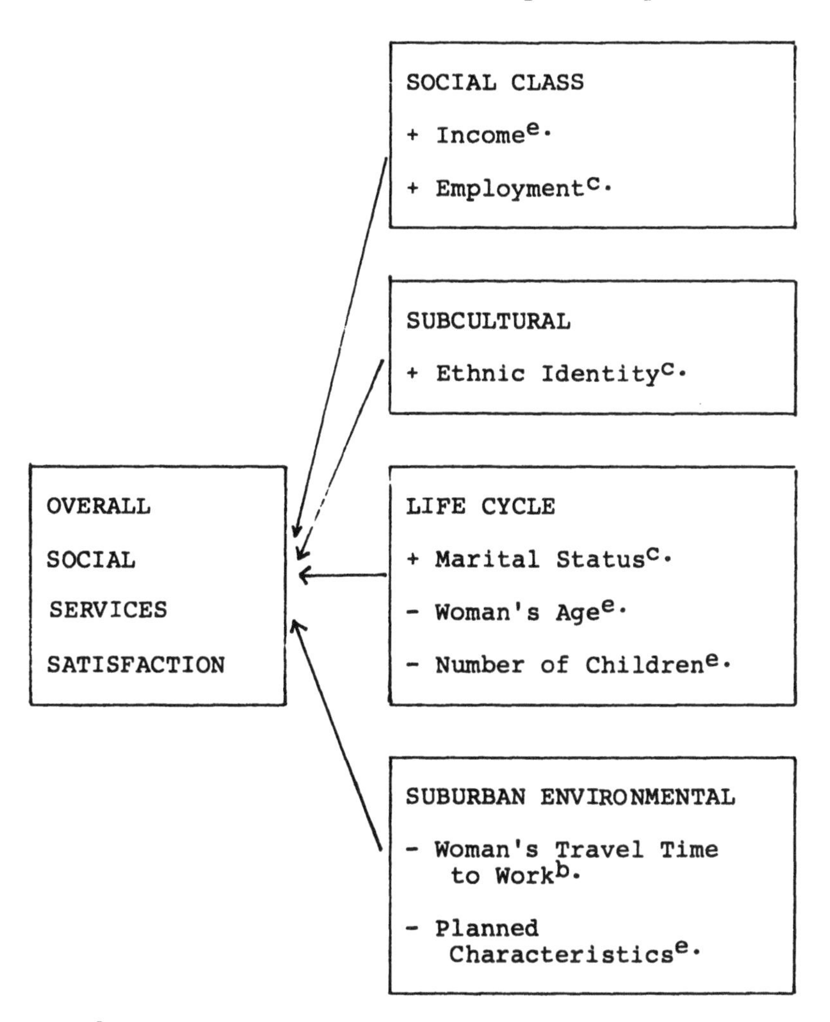


Legend

b. significant at the 0.100 level
c. significant at the 0.050 level
d. significant at the 0.010 level
e. significant at the 0.005 level

FIGURE 5.8 COMPARATIVE INFLUENCE OF INDEPENDENT VARIABLE CATEGORIES ON OVERALL SOCIAL SERVICES SATISFACTION IN ISRAEL

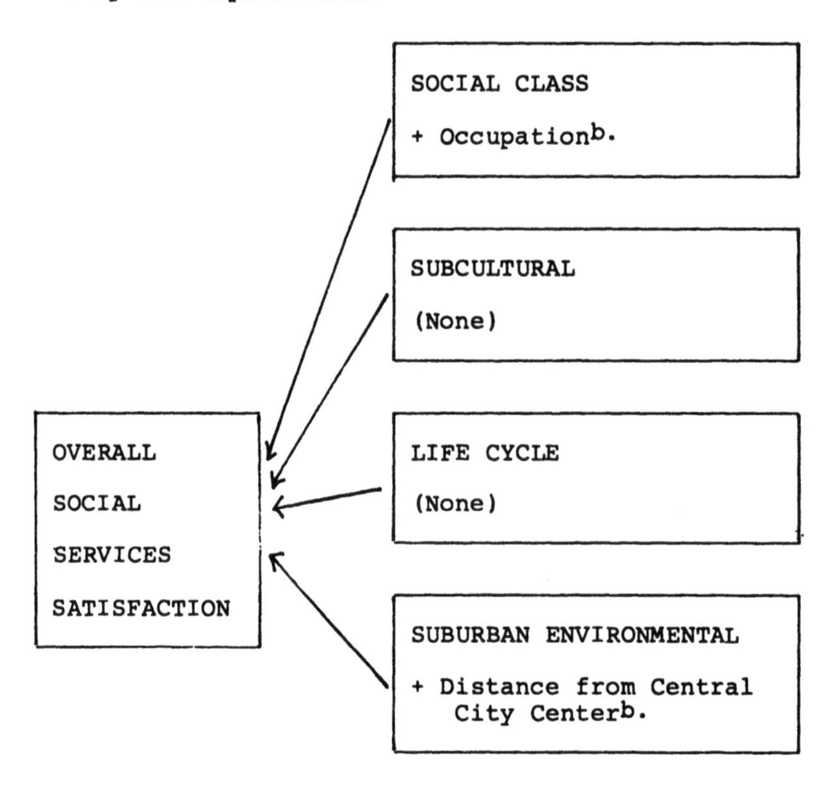


Legend

b. significant at the 0.100 level
c. significant at the 0.050 level
d. significant at the 0.010 level
e. significant at the 0.005 level

FIGURE 5.9 COMPARATIVE INFLUENCE OF INDEPENDENT VARIABLE CATEGORIES ON OVERALL SOCIAL SERVICES SATISFACTION IN THE NETHERLANDS

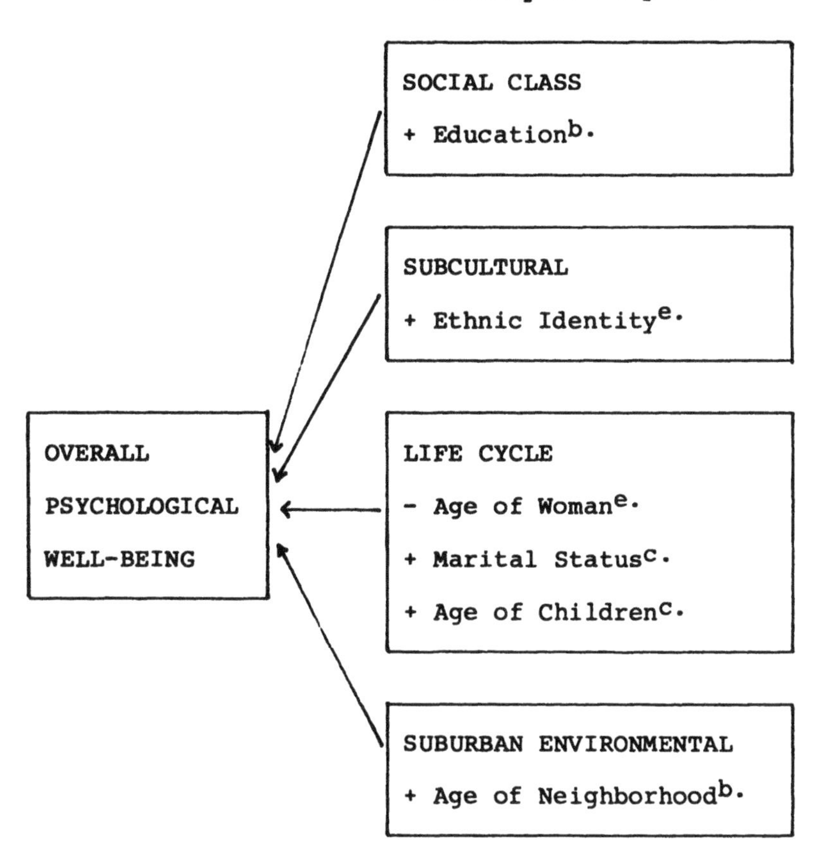


<u>Legend</u>

b. significant at the 0.100 level
c. significant at the 0.050 level
d. significant at the 0.010 level
e. significant at the 0.005 level

FIGURE 5.10 COMPARATIVE INFLUENCE OF INDEPENDENT VARIABLE CATEGORIES ON OVERALL PSYCHOLOGICAL WELL-BEING IN THE UNITED STATES

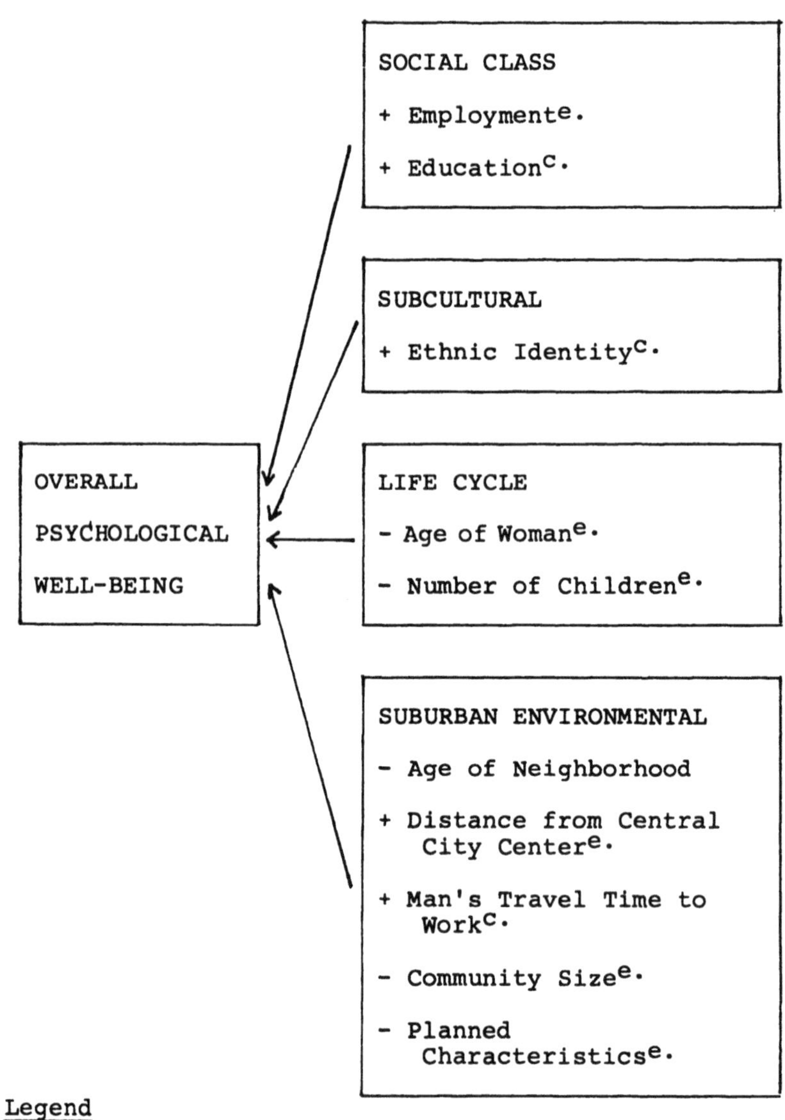


Legend

b. significant at the 0.100 level
c. significant at the 0.050 level
d. significant at the 0.010 level
e. significant at the 0.005 level

FIGURE 5.11 COMPARATIVE INFLUENCE OF INDEPENDENT VARIABLE CATEGORIES ON OVERALL PSYCHOLOGICAL WELL-BEING IN ISRAEL

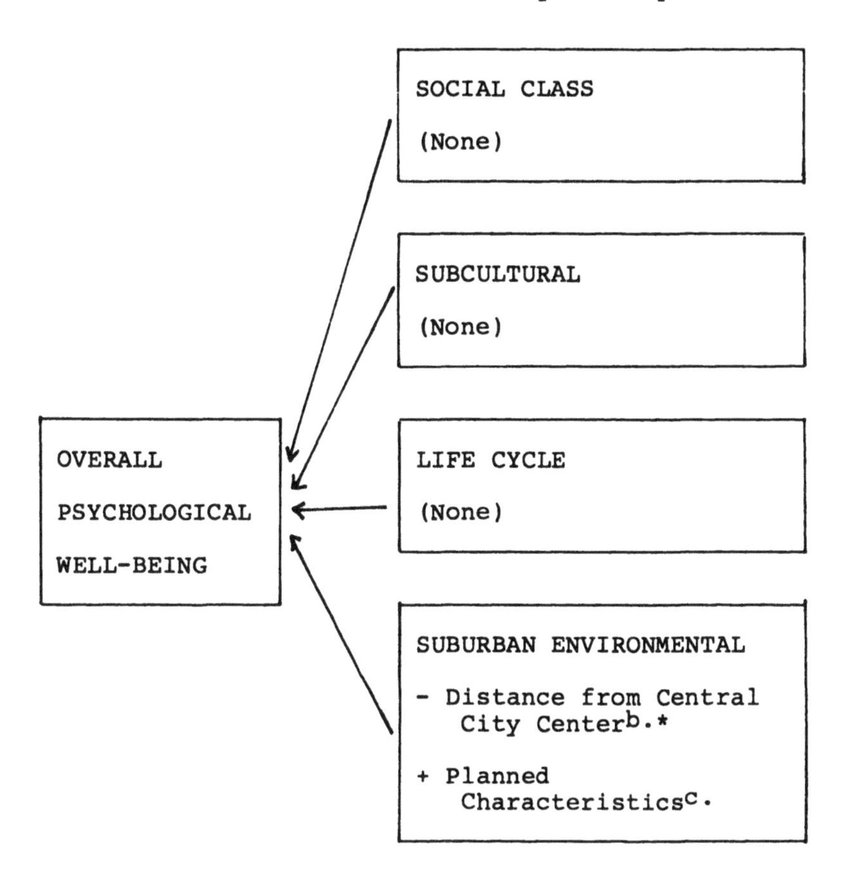


<u>Legend</u>

b. significant at the 0.100 level
c. significant at the 0.050 level
d. significant at the 0.010 level
e. significant at the 0.005 level

* non-working women only

FIGURE 5.12 COMPARATIVE INFLUENCE OF INDEPENDENT VARIABLE CATEGORIES ON OVERALL PSYCHOLOGICAL WELL-BEING IN THE NETHERLANDS

to be those associated with life cycle, and the least important appears to be subcultural influences.

To varying degrees the Israeli and Dutch data approximate the American pattern with one major exception - the influences on psychological well-being. In Israel there is a strong combined impact of social and environmental independent variables, while in the Netherlands only environmental influences affect dependent variable satisfaction.

Thus, for the United States, and to a large extent for the Netherlands, results indicate that the relationships between the theory-related clusters of independent variables and overall satisfaction indexes corroborate the findings of other recent studies.[1] That is, while environmental factors can substantially influence satisfaction with housing and with community services, such factors tend to have a very limited impact on the more global feelings of psychological well-being. For this latter, and perhaps most important aspect of the quality of life, it seems that environmental influences cannot be relied upon. Instead the characteristics of the people themselves should be given most of the explanatory weight.

What do these results suggest for the future of the urban industrialized world?

## B. SOCIAL IMPLICATIONS

Social class influences appear to exert the greatest impact on satisfaction with suburban life in Israel and in the Netherlands, while life cycle factors were most powerful in the United States. In every country subcultural influences were the least important set of social variables.

Among the social class variables, education seemed to have more explanatory power than did income in the United States. Education was also more influential than income with respect to psychological well-being in Israel. The relatively modest impact of income on aspects of quality of life satisfaction should be viewed in the context of the middle class participants in our study for whom it appears that additional increments of income may buy larger and better housing but do not buy more psychological well-being or social services.

We found that a woman's educational level correlated strongly with her satisfaction with community services and social patterns in the United

States and with psychological well-being in the United States and Israel. We do not believe that better educated women are happier because they are culturally advanced. Education implies more than classical intellectual training or refined aesthetic standards. We suspect that educated women are happier in suburbia because they are able to take better advantage of the many opportunities that are available. A person who takes an essentially passsive stance toward the environment will probably find suburbia bland and socially barren when compared to either urban or small town life. An educated person is more likely to see herself in an active role with the power to shape her own experience. Education develops self-knowledge and an awareness of one's one preferences, skills and needs. It provides an understanding of contemporary life and of the social and cultural patterns that may differ from the time and place where one grew up. It develops the intellectual abilities needed for rewarding careers outside the house and the social and interpersonal skills required to reach out to others and initiate friendships, organize groups and also to set limits, resist domination, and to protect one's own need for privacy. An educated person is more likely to know how to obtain information and get results from formal organizations. (She will call the sanitation department to complain about defective storm sewers, and the park and recreation department to find out about programs for her children.) Consequently, community services seem more satisfactory. A well-educated person will seek out interesting classes, workshops, lectures, concerts and art exhibits, many of which are free through public institutions. Just as individuals have to develop skills as consumers, suburban residents must hone their resources in using and profiting from services provided by their communities. Formal education, though not essential to the development of such skills, is certainly useful.

While it is never a surprise when professional educators place a high value on the activity around which they have centered their lives, the data from this study and many others furnish grounds for the particular urgency with which we press the following recommendation.[2] Educational opportunities should be expanded for women at all levels in order to allow them fully to participate in the broad spectrum of societal opportunities. Education for all persons, male and female, should emphasize personal autonomy and develop awareness of the cultural and environmental factors that can lead to fulfilling lives without additional demands for material resources.

As mentioned earlier, of the three sets of social influences, subcultural variables had the least impact on the various indexes of satisfaction. Indeed, in the Netherlands not one subcultural variable had a significant impact on any overall satisfaction score. However, in the United States and in Israel, the emergence of a relationship between a strong sense of ethnic identity and psychological well-being is a potentially important finding that bears further investigation. In industrial societies, where the price of mobility and flexibility may be the loss of social stability, those who have strong ties with their ethnic heritage may have a sense of rootedness that serves an important psychic function. Our findings seem to offer tentative support for those educational and social programs aimed at fostering ethnic pride and identity in a multi-cultural society. Rather than being divisive as some might have feared, the encouragement of a sense of cultural identity appears to create a strong and healthy self-concept. It is important to note that our data refer to persons from many backgrounds who claimed a strong sense of ethnic or cultural identity, whether they are from traditionally oppressed groups such as Blacks and Mexican Americans, whether they belong to more culturally assimilated 'ex-ethnic' groups of European nationalities in the United States, or whether they are part of any of the many cultural, national or religious groups in Israel and in the Netherlands. Programs developing a sense of personal history and pride in heritage should be expanded for all persons as a possible antidote to the homogeneity and anonymity of metropolitan life in the industrial world.

Life cycle variables seem to be the social variables with the greatest influence on satisfaction with urban life in the United States; their importance is substantial in Israel and is of moderate significance in the Netherlands. In each country in this study, marital status was significant and in the United States it appears most often as the variable explaining satisfaction of every kind. The problems of single-headed households will persist for as long as social, religious and political institutions continue to support the supremacy of the nuclear family lifestyle. Whether or not one believes that it is possible or desirable for the single-parent household to be as effective as the two-parent home, one must acknowledge that there will continue to be many divorces and that more women than men will have primary custody of children. Public policy should continue to provide economic support for women and children and most specifically should fund public childcare facilities and adequate public transportation systems. Educational and occupational discrimination against women must be eliminated in order to give the children of households headed by women a fair chance economically. Anti-discrimination legislation, affirmative action policies and changing attitudes have had some impact in recent years, but, as long as women earn considerably less than men in industrialized countries, it is evident that there is much left to be done.[3]

Although not all men and women wish to marry and not all divorced persons wish to remarry, there are many people who prefer a connubial lifestyle and have lacked the skills (or luck) to make such a relationship work. If it is true that married women are substantially happier than unmarried women in a suburban setting, perhaps it would be desirable for public policy to provide greater assistance to married couples who are struggling to maintain their relationships. Rather than coercive economic and tax policies or pious pronouncements from political figures, we would recommend more publicly supported marriage counseling and wider education regarding the factors that lead to successful marriages. Many divorces appear to result from archaic conceptions of marriage relationships. We would in no way encourage any person to get married or to remain married, but we would like to see those who choose a conjugal lifestyle to have as much support as possible in building and maintaining a satisfactory relationship.

Other life cycle variables that have implications for our discussion are a woman's age and the number and ages of her children. One view would suggest that women in their middle years ought to be happy, fulfilled and eager to face a variety of challenging options. To the contrary, we found that age varied negatively with almost all indicators of psychological well-being in the United States and in Israel. Because our sample was limited to women with at least one school-aged child, we cannot compare their satisfaction levels to those with no children or those whose children have left the home. However, other studies, such as that conducted by Campbell, Converse and Rodgers, suggest that older women in all three categories are freed from some of the stresses that accompany the early years of the childrearing cycle.[4] Curiously, our study showed that women's sense of psychological well-being increased as their children got older. Of course, children cannot age without their mother aging as well. On the face of it, this paradoxical finding might seem to argue for bearing children while one is young. In fact, in industrial societies, the trend is in the other direction with testimonials from many women that delaying childrearing has allowed them to achieve educational goals, get started on a career and to establish clearer values about the kind of lives they want to lead.

Usually, women who want to raise a family and pursue a career are forced to delay one or the other. Some women who choose first to establish their careers later decide not to have children. Other women who have children early may never complete their education or enter a career. When the relinquishing of one goal in favor of the other is a true matter of conscious choice, no problem exists. But for many women this is a decision made by default as they assume the roles of wife and mother. Men, of course, may also delay the attainment of personal or professional goals because of demands from the other domain, but the choices here are largely economic. Men are usually not asked to give up several years of their professional growth in order to be parents. At the policy level, then, we recommend that the responsibilities of caring for a home and nurturing children be presented through the media and education as tasks appropriate for any adult. Men and women together can decide which parent should fulfill the primary responsibilities for home care and child care at any point. A national acceptance of this assumption could induce employers to be more open to flexible working schedules such as job sharing, and flexible working arrangements that might change from year to year as family demands changed. Employers might be willing to excuse fathers as well as mothers for occasional parenting responsibilities, such as visiting a teacher or caring for a sick child. This recommendation seems especially relevant in Israel, where the national government encourages large families with children allowance programs and where our data show powerful negative relationships between an increase in the number of children and a woman's overall satisfaction with housing, social patterns and psychological well-being.[5]

Another major policy implication of our findings is that women should be encouraged to widen their options in their middle and later years. For men, the years between thirty and fifty are considered to be the most productive phase of their lives. It is unfortunate that women have been so influenced by the youth culture that some of them consider themselves to be past their peak when their last child starts school. Many universities offer special scholarships and re-entry programs for women who are returning to higher education after a long break in formal schooling. The development of Women's Studies programs helps women to understand their own life experiences in a cultural and historical context and provides models of female achievement. Counseling programs for women and their families can ease the adjustment to new roles for women. Some industries are coming to recognize the special skills and insights of women who have managed homes and handled major responsibilities as community volunteers. In order to minimize the psychological distress implicit in our findings and to prevent an unconscionable waste of talent, we recommend the continuation and expansion of these programs which give financial and moral support to mature women.

## C. ENVIRONMENTAL IMPLICATIONS

This study of suburbia indicates that, with few exceptions, in the United States and in the Netherlands environmental factors have little effect on women's overall sense of happiness or psychological well-being. Perhaps this finding should not be surprising in that behavior has been shown to be more effectively explained by personal characteristics than by environmental variables.[6] Indeed, our findings corroborate those of Zehner in his study of the quality of life in American new communities: 'aspects of the residential environment (including whether or not the area is planned) have only a limited impact on overall life satisfaction'.[7]

While these findings suggest the limitations of environmental influences on the global sense of life satisfaction for a relatively well-off population, it it not clear that the same result would occur with different socio-economic and cultural groups.[8]

For example, our Israeli study indicates a powerful impact of the environment on psychological well-being. Moreover, our study finds that environmental factors are often important influences on the satisfactrion with other quality of life aspects such as housing, community services and social patterns in all three countries. This suggests that environmental planning and design can indeed make important contributions to a convenient and generally supportive context for our lives, but that it may be too much to expect that environmental factors themselves would produce happier lives in all societies.

Our study also reveals the desirability of older inlying single family neighborhoods for women with children in the United States. Apparently in such neighborhoods, the benefits of design and visual familiarity, of accessibility to the metropolitan center, expecially for employment opportunities, and of a supportive, diverse and ongoing social structure can outweigh the costs of lower satisfaction with home maintenance and some public services, such as schools. This finding is similar to what Michelson uncovered about the desirability of single family housing near downtown Toronto.[9] It suggest that governments of American central cities and perhaps of inlying suburbs should prevent the decline of their older neighborhoods through conservation and rehabilitation. This could be pursued not only from the perspective of encouraging home maintenance, such as publicly supported loan programs, but also with an eye to upgrading local public services, particularly schools.[10]

In contrast, more recent Israeli and Dutch outlying areas were generally found to be superior to the inlying older neighborhoods. In Israel, this pattern seemed to be strongly related to the perceived superiority of the quality of housing and community facilities in newer outlying areas, and to the social and psychological rewards of such heterogeneous areas. That is, these relatively new Israeli communities seemed to be more open and socially integrated than older neighborhoods. Thus, newer outlying neighborhoods in Israel appear to reflect that society's policy of residential social integration.[11]

In the Netherlands, not only did housing production cease during World War II, it was slow to resume after 1945. Further, until the early 1960s, standards of physical accommodations conformed to a policy of national austerity aimed at rebuilding the country's infrastructure and industrial base. As a result, amenities in housing were deferred until a national consensus shifted economic policy to a greater emphasis on the consumer sector.[12] Since that time standards have continued to improve, thus rendering newer Dutch neighborhoods generally more desirable housing environments than older areas.

Even with Israeli and Dutch suburban housing densities comparable to those of inlying areas, a pleasant suburban setting previously unavailable to many middle income families in these countries may have become a symbol of social improvement such as that prevalent in the United States a generation before.[13] The danger here is that, without a corresponding decentralization of employment, cultural and transport facilities, new outlying housing environments in these nations will become too socially and physically remote from essential services and activities. At the same time, the Israeli and Dutch governments should continue to upgrade the quality and diversity of housing in inlying neighborhoods in order to make these areas more attractive to middle-income families and thus moderate the costly outward push of the metropolitan population.

Another major finding of our American study is that in relatively new outlying residential areas, planned, higher-density developments such as condominiums and planned unit developments, appear to have more desirable housing and community services than conventional single family neighborhoods for not only women who are single and working, but for all women with young children. These findings were based on the apparent superiority of the high density areas with regard to easier household responsibilities, a greater sense of security, and the accessibility to other community services, such as childcare and usable open space. Since these planned areas offer housing and services which are more supportive and less demanding of women with children than are less-planned areas, they tend to attract a relatively high percentage of women with considerable pressures on their time and energy - working women, especially those who have difficult jobs, and women who are single parents.

As it seems likely that an increasing percentage of American women will be working and thus become more independent in the future, it is probable that the demand for planned, higher density housing environments will increase substantially. It therefore would be useful for American suburban communities to encourage increasing the supply of higher density housing areas in the form of planned neighborhoods with appropriate support services and socializing facilities, as well as the transportation linkages to educational, cultural, employment, and shopping activities. This will not only mean institutional changes, such as re-orienting American zoning ordinances and building codes which often exclude multi-family housing, but also will require broad citizen participation in order to educate the public and the building industry about the positive social, economic, and environmental aspects of such housing as well as about the potential market for its development.[14]

While the findings about the desirability of planned areas were mixed in the Netherlands, such areas in Israel were found clearly inferior to the less-planned neighborhoods. The Israeli findings were primarily related to frugally-designed, publicly planned housing developments characterized by smaller housing units and less accessible sites than those of privately built less-planned areas.[15] Thus, Israel would do well to improve the design and site planning process for public housing developments and encourage an increase in the environmental quality of private sector housing through better infrastructure and related community facilities. Although less of a problem than in Israel, the Netherlands could also improve the appearance and accessibility to community facilities of some neighborhoods.

In all three countries both men's and women's travel time had a similar impact on women's satisfaction with the quality of life. While this does have a male-female social role bias which may lessen in the future, long commutes are still likely to exact a toll on the quality of family life. In addition, it was shown in each society that women's travel time can have substantial negative social and psychological effects. This suggests that efforts should be made to encourage land use and transportation policies which foster a more spatially compact metropolitan area, such as locating jobs closer to housing, in-filling new development on under-utilized parcels, building new residential areas in relatively high densities, and expanding public transportation facilities in order efficiently to integrate regional land use structure. This, in turn, calls for a level of regional planning and management which has heretofore been difficult to attain in the United States due to the decentralized, political structure of metropolitan areas.[16] This, too, will require a considerable amount of public participation and education to inform the diverse metropolitan-wide constitutencies about the wisdom of trading off some local autonomy for the benefits of regional management.[17]

Our study indicates that the residents of smaller political units in all three countries enjoy greater satisfaction with public services than do their central city counterparts. It was also suggested that the differences in satisfaction with these two sets of public service systems were based, in part, on income differences between the average households in the smaller suburban communities and the central city, and that these localized differences create public service inequities as well as great difficulty in regional planning and management of metropolitan areas.

Although some long term forces may be emerging which could narrow the differences between the suburban and central city public services,[18] it seems likely that suburban residents in most societies will try to continue to have control over their local communities and perpetuate central city-suburban inequalities.[19] For this reason there has been pressure within American central cities to decentralize some of the more socially-sensitive public services, such as police patrols and schools, in order to facilitate community control of neighborhoods.[20] We therefore suggest that central cities do whatever they can to upgrade the quality of public services, and wherever feasible, to decentralize authority, or create the 'equivalents of sovereignty' to enable neighborhoods directly to control the management of local public services. Such policies are likely to attract middle and upper-middle class households to the more accessible central city neighborhoods.

Our study questions the desirability of housing environments which characterize much of metropolitan America - new outlying single family tract developments. Indeed, such environments may well be outdated and disfunctional with respect to recent social changes such as those in the status of women.[21] Our study also questions the prudence of the vigorous outward suburban development in metropolitan areas of other industrialized societies, a trend which may generate the same social and physical isolation present in many American and older Israeli suburbs. It remains a question whether resources can keep abreast of such development so that painful trade-offs can be minimized between the benefits of peripheral metropolitan living and those of older and more central locations. As things stand now in most industrial nations, even if these trade-offs must be made, consumers seem to prefer a suburban environment. Thus, some form of regional planning and management system appears to be needed to guide what may be the continuing suburbanization of the industrialized world.

## D. CONCLUSIONS

For centuries nations were easily classified into regions that were clearly urban or rural, or an intermediate category, according to certain definable physical and demographic characteristics. By the mid-1970s, a far greater proportion of the population of industrialized nations lived in suburbs than ever before, but the suburbs had changed from the bedroom communities of an earlier age. As the industrial world becomes more developed and hence more decentralized, cities and small towns have each become suburbanized, and the suburbs, paradoxically, have at once become more like small towns and more urban. This blurring between urban and small town life, or the metropolitanization of the hinterland, has already been noted in the United States,[22] Israel,[23] the Netherlands[24] and many other industrialized nations.[25]

Following the residential expansion outward, shopping facilities, entertainment, services and industry have generated increasing employment opportunities in suburbia.[26] Studies have shown that in the larger suburbs a majority of working people can be employed within their own community.[27] While certain residents continue to commute to the nearby urban center for employment and cultural enrichment, more and more suburban residents center their activities close to home. One can recall Wood's observation of the limited use that suburban dwellers made of central city facilities in the United States.[28] It can be argued that this behavior pattern is reminiscent of the provincialism of small town life. Fischer and Jackson are among the writers who have described suburbia as constellations of small towns.[29]

Our results offer some support for the view

that suburbia functions as a series of small towns. Those people who grew up in small towns seemed to find suburbia most satisfying, suggesting that they found the general social patterns familiar and comfortable. Our respondents who lived in the smaller suburban communities of Los Gatos and Cupertino in the San Jose area, in Raanana and Kefar Sava in the Tel-Aviv region and in Rijswijk near The Hague, were generally more satisfied with the social and community dimensions of their lives than were their central city counterparts. Also highly satisfied were the residents of Willow Glen in San Jose, formerly a separate municipality, which still retains much of the character of a small town within San Jose. Quality of life research shows that small town residents are among the groups expressing the greatest satisfaction with their community attributes.[30] Creating small towns within suburbia, whether the trend takes the form of separate municipalities or just groups of neighborhoods with a central focus, may fill the need of many people for a sense of community and identity. For this reason, the planning of large scale suburban areas should physically reinforce the social concept of neighborhood or of a small unit, a strategy which has long been utilized in the building of new towns in industrialized countries such as England, Israel, Sweden and the Netherlands.[31]

Even though the social patterns of suburbia tend toward small town orientation, the character of the population seems to be becoming more urbanized. While the traditional suburbs were dominated by nuclear families in the childrearing years of the life cycle, our research and review of literature show that contemporary suburbia attracts smaller households of single adults, divorced parents, childless couples, senior citizens and many other groups that characterize the heterogeneity of urban areas.[32] These are the groups that do not seem to partake of the small town aspects of suburbia. Specifically, working women, single women and women with strong ethnic ties report that they do much of their socializing outside their neighborhoods. Thus, members of these non-traditional groups often live in relative isolation from their immediate neighbors and seek friendship and group participation via non-spatial routes.[33]

Yet, there is no need for conflict between those groups who wish to treat suburbia as a small town environment and those who wish to experience its more urban aspects. Those areas between city and country can be envisioned as a checkerboard pattern of small town centers whose facilities are shared by residents of both high- and low-density housing, of traditional and non-traditional lifestyles. As mentioned earlier, in terms of public policy, it is important that the needs of the urban suburbanite not be subordinated to those of small town preferences. Therefore, combined high- and low-density housing should be encouraged in suburban areas, and public transportation should be expanded and not neglected in favor of the automobile.

Many of our proposals for the future of urban development in the industrialized world are not new. For decades others have argued in favor of regional planning for economic, social, political and environmental reasons.[34] Indeed, as housing and commutation costs escalate, pressures for cost-reducing and energy efficient higher-density housing and for more public transportation are likely to mount anyway. What we have to add to this discussion about metropolitan areas is that these planning proposals are desirable for still other very important reasons - for industrialized societies to facilitate equal opportunities for women and non-traditional suburban residents, and to accommodate the social changes related to this process.

NOTES

[1] Robert B. Zehner, Indicators of the Quality of Life in New Communities (Ballinger, Cambridge, Mass., 1977), pp.11-19; and Donald N. Rothblatt, Daniel J. Garr, and Jo Sprague, The Suburban Environment and Women (Praeger Publishers, New York, 1979), Chapter 4.

[2] For another study with positive findings about the educational benefits for women, see Angus Campbell, Philip E. Converse and William Rodgers, The Quality of American Life: Perceptions, Evaluations, and Satisfactions (Russell Sage Foundation, New York, 1976), pp.424-9.

[3] For example, see Lindsy Van Gelder, 'Strains of "Oh Promise Me" Emanating from the White House', San Jose News (February 8, 1979), p.10C; and Israel Central Bureau of Statistics, Statistical Abstract of Israel: 1979 (Jerusalem, 1980), p.358.

[4] Campbell, Converse and Rodgers, The Quality of American Life, p.405.

[5] Joseph Neipris, Social Welfare and Social Services in Israel: Policies, Programs, and Current

Issues (The Hebrew University of Jerusalem, Jerusalem, 1981), Chapter 3.

[6] Claude S. Fischer, To Dwell Among Friends: Personal Networks in Town and City (University of Chicago Press, Chicago, 1982): Claude S. Fischer and Robert M. Jackson, 'Suburbs, Networks, and Attitudes' in B. Schwartz (ed)., The Changing Face of Suburbs (University of Chicago Press, Chicago, 1976), pp.279-309.

[7] Zehner, Indicators of the Quality of Life, p.15.

[8] Donald N. Rothblatt, 'Improving the Design of Urban Housing' in Vasily Kouskoulan (ed.), Urban Housing (National Science Foundation, Detroit, 1973), pp.149-54.

[9] William W. Michelson, Man and His Urban Environment (Addison-Wesley, Reading, Mass., 1975).

[10] Richard E. Starr, 'Infill Development - Opportunity of Mirage', Urban Land, vol.39, (1980), pp.3-5.

[11] Naomi Carmon and Bilha Mannheim, 'Housing as a Tool of Social Policy', Social Forces, vol.58, no.1, (1979), pp.336-51.

[12] Jan Van Weesep, 'Intervention in the Netherlands: Urban Housing Policy and Market Response', Urban Affairs Quarterly, vol.19, no.3 (1984), pp.329-53.

[13] Aviva Lev-Ari, 'Spatial Behavior and Overt Residential Preference', unpublished Master's Thesis, Hebrew University of Jerusalem, 1978.

[14] Paul Davidoff and Mary Brooks, 'Zoning Out the Poor', in P. C. Dolce (ed.), Suburbia: The American Dream and Dilemma (Anchor Press, New York, 1976), pp.135-66.

[15] Eli Borukhov, Yona Ginsberg, and Elia Werczberger, 'Housing Prices and Housing Preferences in Israel', Urban Studies, vol.15, no.2 (1978), pp.187-200.

[16] Richard L. Bish and Hugh O. Nourse, Urban Economics and Policy Analysis (McGraw-Hill, New York, 1975), pp.204-7.

[17] Donald N. Rothblatt, Regional Planning: The Multiple Advocacy Approach (Praeger Publishers, New York, 1982), Chapter 8.

[18] Efforts to equalize the quality of local public services is demonstrated by the 1971 California Supreme Court Decision, Serrano vs. Priest, which requires the equalization of available financial support per pupil in school districts throughout the state. Despite this trend, there is evidence which suggests that citizens may not be

aware of qualitative differences in their local public services. See Brian Stipak, 'Citizen Satisfaction with Urban Services: Potential Misuse as a Performance Indicator', Public Administration Review, vol.39, no.1 (1979), pp.46-52.

[19] John R. Logan and Mark Schneider, 'Governmental Organization and City/Suburb Income Inequality, 1960-1970', Urban Affairs Quarterly, vol.1 , no.3 (1982), pp.303-18.

[20] Bruce London, Donald S. Bradley, and James R. Hudson, 'The Revitilization of Inner-City Neighborhoods', Urban Affairs Quarterly, vol.15, no.4 (1980), pp.373-80.

[21] Susan Saegert, 'Masculine Cities and Feminized Suburbs: Polarized Ideas, Contradictory Realities', Signs, vol.5, Spring Supplement (1980), pp.96-111.

[22] John F.Long Population Deconcentration in the United States (U.S. Bureau of the Census, Washington, D.C., 1981), Chapters 2 and 3.

[23] Alexander Berler, et al., Urban-Rural Relations in Israel: Social and Economic Aspects (Settlement Study Centre, Rehovot, Israel, 1970), pp.89-90.

[24] Francois J. Gay, 'Benelux' in Hugh D. Clout (ed.), Regional Development in Western Europe (John Wiley and Sons, New York, 1981), pp.179-209.

[25] Gordon E. Cherry, 'Britain and the Metropolis: Urban Change and Planning in Perspective', Town Planning Review, vol.55, no.2 (1982), pp.5-33; and Peter Hall and Denis Hay, Growth Centers in the European Urban System (University of California Press, Berkeley, California, 1980), Chapters 1 and 7.

[26] Hall and Hay, Ibid., pp.87-9.

[27] Stanley Buder, 'The Future of the American Suburbs', in Dolce (ed.), Suburbia: The American Dream and Dilemma, p.205; and Cherry, 'Britain and the Metropolis', pp.27-32.

[28] Robert C. Wood, Suburbia: Its People and Their Politics (Houghton-Mifflin, Boston, 1958), p.107.

[29] Fischer and Jackson, 'Suburban Networks and Attitudes', pp.299-304.

[30] Campbell, Converse and Rodgers, The Quality of American Life, p.236 .

[31] For example, see: Artur Glikson, 'Urban Design in New Towns and Neighborhoods', Ministry of Housing Quarterly, (December 1967), pp.47-51; David Poponoe, The Suburban Environment: Sweden and the United States (The University of Chicago Press,

Chicago, 1977), pp.207-15; Donald N. Rothblatt, 'Housing and Human Needs', Town Planning Review, vol.42, no.2, (1971), pp.130-44.

[32] The overall decline in the average number of persons per household due to the increase of such non-traditional groups has been well documented in industrialized societies. See for example, Israel Central Bureau of Statistics, Statistical Abstract, p.63; Jan Van Weesep, 'Intervention in the Netherlands; Urban Housing Policy and Market Response', p.344.

[33] The relative degree of social isolation of these groups is similar to the less intimate social relations of the 'community of limited liability' compared to that of the 'urban village' described in Avery M. Guest and Barrett A. Lee, 'The Social Organization of Local Areas, Urban Affairs Quarterly, vol.19, no.2 (1983), pp.217-40.

[34] See for example, the work of Luther Gulick, 'Metropolitan Organization', Annals of the American Academy of Political and Social Science, vol.314 (1957), pp.57-65; and Dolores Hayden, Redesigning the American Dream: The Future of Housing, Work and Family Life (W. W. Norton, New York, 1984).

# CHAPTER 6

# IMPLICATIONS FOR CROSS-NATIONAL RESEARCH

The history of the social sciences has been marked by a fascination to compare human collectivities and from them distill generalizations based on inductive, empirical evidence. From the classical antiquity of Aristotle to Ibn Khaldun's fourteenth century Kitab al-Ibar (universal history), to the more modern efforts of Montesquieu, Marx, Spencer, Durkheim and Weber, comparative inquiry today is built on a well-seasoned foundation of precedent.[1] However, only since World War II have comparative studies achieved prominence within the social sciences. This late blooming cannot be attributed to the indifference of researchers, but rather to the inherent difficulties which the conduct of such studies poses.

While large amounts of data in a variety of fields have long been available, they have been subject to the control of government bureaucracies which have been resistant to the needs of academics. Therefore, circumventing official indifference and structuring inquiries along suitable lines has long been a problematic but essential priority for scholarly researchers. About 1945 a new breed of academic entrepreneur arose, equipped with both a data base, obtained through survey research, and the intellectual control of a project so that a broad spectrum of information could be obtained which would be beyond the scope of what had previously been the domain of often inaccessible official statistics.[2]

The first of the post-1945 cross-national surveys was the work of Hadley Cantril and William Buchanan, How Nations See Each Other, a project coordinated by UNESCO and published in 1953.[3] Given the nature of things in the early years of postwar Europe, most of the major cross-national surveys of this period were dominated by American scholars with access to the major funding necessary to conduct such complex projects. However, with the establishment of the Vienna Centre by the International Social Science Council in 1963, survey research not only became internationalized, but further served as a base for cooperation between the East and West Blocs. In its first decade, the Centre funded cross-national research in such diverse areas as time budgets, tourism, images of the world in the year 2000, automation, and juvenile delinquency, with each study utilizing data from both sides of Europe's two ideological camps.[4]

Stefan Nowak, a Polish participant in the work of the Vienna Centre, has classified cross-national surveys into two major groupings: those which are nation-oriented and those which are variable-oriented.[5] It is not the primary purpose of the study discussed in the preceding chapters to compare nations or to focus on particular institutions in their societies. Rather, as a variable-oriented undertaking, the study of suburbia in three industrialized nations provides the opportunity to verify and explore several conceptualizations. By utilizing four theory-related clusters of universally defined independent variables (social class, subcultural, life cycle and environmental) and observing their influence on a set of dependent quality of life variables (satisfaction with housing environment, community services, social patterns, and psychological well-being) a beginning can be made on the development of further insights concerning metropolitan life. Further it is possible to suggest directions for policy formulations based on equivalences that define the United States, Israel and the Netherlands, three advanced, industrial, democratic, and primarily capitalist nations.

As suggested by Nowak, we have endeavored to replicate anticipated findings from the San Jose, California metropolitan area and, by obtaining data from samples in comparable sub-national units - metropolitan areas in Israel (Tel-Aviv and Jerusalem) and the Netherlands (The Hague agglomeration) - it has been possible to test the universality of the expectations posited.[6] The basis for comparison lies in the universality of metropolitan decentralization, one of the most salient determinants of human activity patterns since 1945.

This would be an appropriate juncture to

discuss whether these three nations are in fact sufficiently similar to make comparisons among them meaningful. If one were to assess characteristics such as history, geography, relative influence in the world politically, economically, or militarily, or even more immediate factors such as how land uses are allocated and controlled, it is possible that dissimilarities in these societies might outweigh equivalences.[7] However, Nowak has cautioned that phenomena which are relationally similar do not necessarily have to be identical insofar as their absolute properties are concerned. Indeed, relationships across countries are never entirely symmetrical; it is therefore the researcher's responsibility to establish equivalence.[8]

As shown in Chapter 1, the problems of cities and the allure of their suburban rings are a recurring theme in modern industrialized nations. It can be argued that this study of suburbia in the United States, Israel and the Netherlands is indeed characterized by three types of equivalence identified by Nowak.[9] By cultural equivalence, it is meant that the phenomenon at hand, metropolitan decentralization, is viewed in a similar manner, regardless of specific cultural differences. Contextual equivalence is also applicable in that the areas selected for study are part of metropolitan agglomerations, a higher level aggregate with similar properties. And, by serving a comparable role in the functioning of each national system, relational equivalence further defines the components of this study.

A second issue which has been raised when the notions of comparability or equivalence surface is the relative impact of human constants. Are there, as Grimshaw has queried, certain regularities in social behavior that are universal as opposed to those which are system-specific?[10] In research conducted in Mexico, Poland, Australia, and Argentina, Kevin Lynch has observed that, despite disparate environmental and institutional conditions, there are human constants in the way children utilize their world.[11] Therefore, how similar are the systems which are under discussion and, if they are similar, how does one evaluate their influence on behavior as opposed to that exercised by more universal forces?

The selection of nations for this tri-national study was inevitably a pragmatic decision for such is a necessary ingredient in any large scale research effort. However, as noted in Chapter 2, in choosing research sites where capitalist economies predominate and where traditions of stable, democratic government are in place, a theoretical imperative emerges. In order to strengthen this aspect of the study it becomes necessary to demonstrate that considerations of comparability at the national level were the chief determinants of the organization of this project. Therefore, by utilizing data which have been collected for the World Handbook of Social and Political Indicators, it is possible to compare measures pertaining to (a) the size of government and the allocation of resources; (b) national economic structure; and (c) wealth and well-being, so that a richer appreciation of the particular contexts of metropolitan decentralization in all three nations can be achieved.[12]

As shown in Table 6.1, the size of government and its allocation of resources offer useful bases for comparison and contrast among the three nations within the distribution of more than 130 contained within the World Handbook's rankings. All three nations rank reasonably closely when measures of general government expenditures are considered. Although the percentage of government expenditures is considerably less in the United States, it is comparable with other nations such as France and West Germany; similar data were not available from East Bloc states. A similar pattern prevails with general government revenues but, when central government data are considered, the United States ranks far behind Israel and the Netherlands. This discrepancy can be easily explained by the far greater role played by state and multiplicities of local governments in the American federal system. Each country spends proportionately similar amounts on education and public health, but Israel's large military budget has continued to distort its economy. This accounts for the disproportionately large share of government consumption and for the large allocations of military manpower in Israel. Conversely, the United States and the Netherlands are remarkably similar on these measures and all three nations remain on virtually the same level when the impacts of private consumptions and gross domestic investment are considered.

Levels of economic development must also be considered in questions of equivalence among nations. As shown in Table 6.2, the structure of the three nations' economies is quite similar. Although all three industrialized at different points in

TABLE 6.1 COMPARATIVE MEASURES OF THE SIZE OF GOVERNMENT AND THE ALLOCATION OF RESOURCES: U.S.A., ISRAEL AND THE NETHERLANDS

| Attribute | USA | Date | Rank | Israel | Date | Rank | Netherlands | Date | Rank |
|---|---|---|---|---|---|---|---|---|---|
| General Govt. Expenditure as % of GDP[a]. | 30.6 | 1973 | 19 | 46.9 | 1972 | 1 | 43.5 | 1973 | 2 |
| General Govt. Revenue as % of GDP | 32.3 | 1973 | 23 | 38.7 | 1972 | 11 | 49.0 | 1973 | 2 |
| Central Govt. Expenditure as % of GNP[b]. | 20.0 | 1979 | 53 | 64.0 | 1978 | 1 | 51.0 | 1980 | 4 |
| Central Govt. Revenue as % of GDP | 21.0 | 1979 | 62 | 53.0 | 1978 | 5 | 52.0 | 1980 | 7 |
| Military Expenditure as % of GNP | 5.1 | 1978 | 37 | 24.3 | 1978 | 1 | 3.3 | 1978 | 64 |
| Public Education Expenditure as % of GNP | 6.0 | 1978 | 30 | 8.5 | 1977 | 5 | 8.4 | 1977 | 6 |
| Public Health Expenditure as % of GNP | 3.7 | 1978 | 22 | 4.7 | 1977 | 14 | 0.3 | 1978 | 119 |
| Military Manpower per 1,000 Working Age Persons | 15.5 | 1975 | 36 | 93.4 | 1975 | 1 | 13.0 | 1975 | 45 |
| General Govt. Consumption as % of GDP | 18.0 | 1978 | 29 | 36.0 | 1978 | 1 | 18.0 | 1978 | 29 |
| Private Consumption as % of GDP | 64.0 | 1978 | 60 | 58.0 | 1978 | 81 | 59.0 | 1978 | 78 |
| Gross Domestic Investment as % of GDP | 18.0 | 1978 | 84 | 23.0 | 1978 | 51 | 21.0 | 1978 | 67 |

a. Gross Domestic Product b. Gross National Product

Source: Charles Lewis Taylor and David A. Jodice, World Handbook of Political and Social Indicators, 3rd ed., 2 vols. (Yale University Press, New Haven and London, 1983), I, tables 1.1-1.4, 1.7-1.9, 1.11-1.14

TABLE 6.2 COMPARATIVE MEASURES OF NATIONAL ECONOMIC STRUCTURE: U.S.A., ISRAEL AND THE NETHERLANDS

| Attribute | USA | Date | Rank | Israel | Date | Rank | Netherlands | Date | Rank |
|---|---|---|---|---|---|---|---|---|---|
| % of Labor Force in Agriculture | 3.0 | 1977 | 120 | 8.0 | 1977 | 111 | 6.0 | 1977 | 114 |
| % of Labor Force in Industry | 33.0 | 1977 | 31 | 38.0 | 1977 | 21 | 45.0 | 1977 | 9 |
| % of Labor Force in Services | 64.0 | 1977 | 3 | 55.0 | 1977 | 12 | 49.0 | 1977 | 22 |
| % of GDP Accounted for by Agriculture | 3.0 | 1978 | 112 | 6.0 | 1978 | 98 | 5.0 | 1976 | 103 |
| % of GDP Accounted for by Industry | 29.0 | 1978 | 42 | 27.0 | 1978 | 53 | 30.0 | 1976 | 37 |
| % of GDP Accounted for by Services | 67.0 | 1978 | 5 | 67.0 | 1978 | 5 | 57.0 | 1976 | 31 |

Source: Charles Lewis Taylor and David A. Jodice, World Handbook of Political and Social Indicators, 3rd. ed., 2 vols. (Yale University Press, New Haven and London, 1983), I, tables 6.1-6.6

history - the United States during the early and mid-nineteenth century; the Netherlands at the end of the nineteenth century; and Israel still more recently - they are all in a transition from an industrial to a post-industrial, service-based economy. Further, all three, despite their role as exporters of agricultural products - the United States with grains, Israel with citrus, and the Netherlands with its horticultural sector - nevertheless have low levels of employment and low percentages of their Gross Domestic Products derived from agriculture. Therefore, all three countries have experienced comparable levels of development and are thus subject to similar land use pressures brought on by such an economic structure.

A third approach to national equivalence may be found in measures of wealth and well-being (see Table 6.3). The resources of a nation and how they are allocated speak directly to measures of standard of living and quality of life. Although per capita figures may be misleading, there is as yet no widely-accepted measure of the distribution of consumption.[13] Even so, it is clear that all three nations rank high on these measures and, while the United States dominates in a number of areas, Israel and the Netherlands also enjoy relatively high standards of living. There are areas in which Israel does lag behind the other two nations, such as Gross National Product per Capita, Energy Consumption per Capita, and the rates of television and telephone ownership. However, with high rates of Protein per Capita per Diem, Physicians per Million Population, and enrollment in higher education, the Israeli government is actively taking measures to increase standards of living, a formidable task given that nation's extensive post-1948 absorption of immigrants. Other research has confirmed that Israel's perceived quality of life is comparable with that of other industrialized nations, such as the United States, Japan, and West Germany (see Table 6.4).

Having made a case for essential comparability among these nations, there are questions of an ecological nature that warrant discussion. What is meant by 'suburbia' or by 'metropolitan'? Schwirian has argued that there is more agreement among nations as to what constitutes a 'city' than there is for how a metropolitan area is to be defined.[14] Some agglomerations may contain significant populations in low-density, quasi-rural areas while others, such as those in the Netherlands, are compact

TABLE 6.3 COMPARATIVE MEASURES OF WEALTH AND WELL-BEING: U.S.A., ISRAEL AND THE NETHERLANDS

| Attribute | USA | Date | Rank | Israel | Date | Rank | Netherlands | Date | Rank |
|---|---|---|---|---|---|---|---|---|---|
| GNP Per Capita (in 1,000 US $) | 9.77 | 1978 | 9 | 3.73 | 1978 | 30 | 9.20 | 1978 | 13 |
| Energy Consumption per Capita (in Kilograms) | 10,999 | 1975 | 5 | 2,769 | 1975 | 37 | 6,027 | 1975 | 12 |
| Calories per Capita per Diem | 3,576 | 1977 | 6 | 3,141 | 1977 | 30 | 3,338 | 1977 | 24 |
| Protein per Capita per Diem | 104.3 | 1974 | 7 | 101.5 | 1974 | 10 | 87.6 | 1974 | 31 |
| Physicians per Million Population | 1,629 | 1976 | 22 | 2,676 | 1973 | 3 | 1,605 | 1974 | 23 |
| Infant Mortality Rate per 1,000 Live Births | 16 | 1975 | 111 | 22 | 1975 | 101 | 11 | 1975 | 123 |
| School Enrollment at the Primary and Secondary Levels as a % of School-Age Population | 87 | 1977 | 32 | 88 | 1975 | 30 | 98 | 1977 | 3 |
| Enrollment in Higher Education per Million Population (in 1,000s) | 52.2 | 1975 | 1 | 22.0 | 1975 | 7 | 21.1 | 1975 | 9 |
| Televisions per 1,000 Population | 571 | 1975 | 1 | 141 | 1975 | 37 | 267 | 1975 | 11 |
| Telephones per 1,000 Population | 697 | 1975 | 1 | 238 | 1975 | 22 | 371 | 1975 | 13 |

Source: Charles Lewis Taylor and David A. Jodice, World Handbook of Political and Social Indicators, 3rd. ed., 2 vols. (Yale University Press, New Haven and London, 1983), I, tables 3.6, 3.7, 4.4-4.6, 4.8, 5.1, 5.2, 5.7, 5.9

TABLE 6.4 COMPARATIVE LEVEL OF SOCIOECONOMIC DEVELOPMENT AND PERCEIVED QUALITY OF LIFE: 1965

| Country | Developmental Index | Mean self-ratings on ladder scale |
|---|---|---|
| USA | 1.00 | 6.6 |
| West Germany | 0.71 | 5.3 |
| Israel | 0.67 | 5.3 |
| Japan | 0.60 | 5.2 |
| Poland | 0.45 | 4.4 |
| Cuba | 0.35 | 6.4 |
| Panama | 0.31 | 4.8 |
| Yugoslavia | 0.19 | 5.0 |
| Philippines | 0.17 | 4.9 |
| Brazil | 0.16 | 4.6 |
| Dominican Republic | 0.16 | 1.6 |
| Egypt | 0.14 | 5.5 |
| Nigeria | 0.02 | 4.8 |
| India | 0.00 | 3.7 |

Source: Hadley Cantril, The Pattern of Human Concerns (Rutgers University Press, New Brunswick, N.J., 1965), Table XII:1

and well-defined. Such an inconsistency is not an issue in this study.

On a less aggregated level, are the sub-areas chosen for study in the three nations comparable? Would an idealized individual, with years of experience in all three societies, have selected the same neighborhoods as those which were utilized in this survey? As Schwirian has recognized, there are problematic variations in the type, scale and quality of sub-area data sufficient to frustrate any researcher seeking to extract cross-national equivalence.[15] If problems with sources of data preclude a choice of study area based on rigorous quantitative comparison, what alternatives can be relied upon and how might they impinge on considerations of equivalence? This study relied on the opinions of scholars and local government officials to produce subjective equivalence, that is, to identify representative residential areas in a social and economic sense on a continuum of distance from the metropolitan center. Informational exchanges aimed at this end can be a most enlightening aspect of study in another society.

Yet another ecological issue is the interregional character of metropolitan equivalence. Are comparisons valid between areas which are in different stages of growth? Can equivalent sub-areas be identifed given the outward dissimilarities of some metropolitan agglomerations? In this study, all three regions are still experiencing decentralization; however, sunbelt vs. frostbelt dynamics in the United States might raise theoretical impasses.

Still a fourth ecological issue is the question of mobility. This has been a public policy outcome paramount in American metropolitan decentralization and has emphasized the social and fiscal inequalities between central cities and their suburban rings. As a major outcome of public policies, its differential ecological implications underscore the failure of urban ecologists to include policy considerations in their descriptions of the dynamics of cities.[16] Mobility may be thought of as a three-pronged issue. First, as Lucy has suggested, societies which vest more resources in the private sector might expect mobility to be both the cause and effect of growing divergences between the suburban ring and the central city.[17] Thus, centrifugal movement may be more pronounced in economies where the market process allows the consumer greater choice, especially with respect to housing. Such is most obvious in the United States, while in the Netherlands, for example, there are not sufficient incentives to foster mobility in the housing market among certain groups.

Second, cultural and historical factors may offset the attractions of suburbs so that central cities may be viewed as both valued artifacts and dominant components of a national tapestry. According to Lucy, 'the greater the social, cultural and historic attachments of people to cities, the more the resulting political culture will support public policies which tend to stabilize and refurbish the central cities in metropolitan areas'.[18] Such an influence would certainly be expected in The Hague, the Dutch political seat of government, or in Jerusalem, with its timeless importance for three of the world's major religions. As a focal point for these concerns, migration from the central city would not be expected on the same scale or in the same context in Israel or in the Netherlands as in the United States.

A third facet of the mobility question is the institutional opportunities for metropolitan decentralization. Low density sprawl is restricted by government policies in Europe as is the autonomy which can be exercised by units of local government.[19] This stands in contrast to the highly influential and varied repertory of local government powers in the United States which control everything from land use to primary and secondary education.[20]

Logically, if the above holds true, one might expect a greater range of environmental satisfaction in the United States and a more homogeneous set of scores in the Netherlands, with perhaps Israel occupying the middle ground. Instead, when one examines data in Chapter 4 concerned with neighborhood location (see tables 4.22-24), it can be seen that inlying residential neighborhoods in San Jose, California, have not lost their desirability when compared to more distant areas closer to the metropolitan fringe. Respondents' preferences for this type of neighborhood prevail across a broad range of dependent variables. Scores from the Netherlands and from Israel follow what one might have imagined to be the American pattern, emphasizing the psychological attributes of a lifestyle removed from the central city.

That preference is stated with even more strength when, housing aside, respondents are asked to weigh the desirability of smaller suburban localities vis-a-vis the larger central city. Despite consumer preferences for more centrally located suburban housing in San Jose, the quality of community services in its suburbs is preferred quite decisively; such is also the case in Israel (see Tables 4.28-30). This pattern holds on a more modest scale in the Netherlands; while housing did not emerge as an issue in suburbs vs. city in the United States and in Israel, it did in the Netherlands, chiefly as an outgrowth of the higher standards of newer dwelling units, most of which were built on relatively outlying land developed since the mid-1960s.

Perhaps, ultimately, small is beautiful and this proposition constitutes a regularity in social behavior that is universal, despite the properties of any social system, be it uncontrolled capitalist urban sprawl, the centralized planned environment of the East Bloc, or the political consensus of quasi-capitalist nations of western Europe. As this research has shown, it is a desirable goal to create the ambience of small towns within our metropolitan areas; but how can this be achieved, and are there adequate sources of information that can steer us in that direction? Ward has questioned whether existing measures of citizen satisfaction are adequate, a reasonable question given the nature of the highly aggregated data she relied upon in her inquiry into quality of life attitudes in urban regions of western Europe.[21] One problem encountered with such data is the divergent responses when individuals are polled from a wide range of economic and cultural backgrounds. This tri-national study has controlled for economic and social status to a large degree and has selected respondents from a range of environments which share what is believed to be a high degree of equivalence.

A second problem noted by Ward addresses the choice of criteria for adequately assessing the quality of life. As discussed in Chapter 3, the choice of dependent and independent variables utilized in this study, and the scaled responses designed to yield measures of satisfaction have been widely employed in the past. How universal are they in applicability? Judging from the findings of this tri-national survey, patterns have emerged which suggest consistency in response patterns from all three countries. Yet, despite the limitations imposed by Ward's data, she, too, calls for measures which will transfer 'the positive qualities of the less populated communities... to the large urban complexes'.[22]

A third problem in comparative cross-national research is the issue of how societies differ in their receipt or enjoyment of various goods.[23] Consider, for example, the question of community services in the three nations studied. As shown in Table 6.5, satisfaction with such services is highest in the Netherlands, emphasizing parks, schools, and transportation, all of which are community goods. In contrast, entertainment received a higher satisfaction score in the United States, activities which may be purchased in individual increments as more market-related services. The only community service for which the Israeli sample registered the highest level of satisfaction was for security, a surprising choice given the continuing unrest in the Middle East.

If the Israeli population has been subject to more terrorist activity than that of the other two nations, one might anticipate feelings of insecurity to be registered by that sample. Yet, precisely the reverse occurred. Is it because extensive security

TABLE 6.5 COMPARATIVE DEPENDENT VARIABLE MEAN SCORES: U.S.A., ISRAEL AND THE NETHERLANDS

| Satisfaction Index | Satisfaction Scores[a.] | | | | | |
|---|---|---|---|---|---|---|
| | U.S.A./ISRAEL | | U.S.A./NETHERLANDS | | ISRAEL/NETHERLANDS | |
| a. Housing Environment | 5.59[e.] | 4.80 | 5.59[e.] | 5.12 | 4.80 | 5.12[e.] |
| 1. Privacy | 5.82[e.] | 5.39 | 5.82[e.] | 4.92 | 5.39[e.] | 4.92 |
| 2. House & Lot | 5.56[d.] | 5.09 | 5.56[b.] | 5.49 | 5.09 | 5.49[e.] |
| 3. Household Responsibilities | 5.22[e.] | 4.20 | 5.22[e.] | 4.87 | 4.20 | 4.87[e.] |
| 4. Appearance | 5.77[e.] | 4.57 | 5.77[e.] | 5.17 | 4.57 | 5.17[e.] |
| b. Community Services | 5.10[e.] | 4.09 | 5.10 | 5.12 | 4.09 | 5.12[e.] |
| 1. Parks | 4.97[e.] | 4.48 | 4.97 | 5.34[e.] | 4.48 | 5.34[e.] |
| 2. Schools | 5.18[e.] | 4.83 | 5.18 | 5.79[e.] | 4.83 | 5.79[e.] |
| 3. Security | 5.24 | 5.14 | 5.24[e.] | 4.32 | 5.14[e.] | 4.32 |
| 4. Child Care | 4.18[e.] | 3.77 | 4.18 | 4.24 | 3.77 | 4.24[d.] |
| 5. Transportation | 5.42[e.] | 3.70 | 5.42 | 6.01[e.] | 3.70 | 6.01[e.] |
| 6. Entertainment | 5.49[e.] | 2.93 | 5.49[e.] | 5.02 | 2.93 | 5.02[e.] |
| c. Social Patterns | 5.33[e.] | 4.42 | 5.33[e.] | 4.71 | 4.42 | 4.71[d.] |
| 1. Friendships | 5.66[e.] | 4.78 | 5.66[e.] | 5.25 | 4.78 | 5.25[e.] |
| 2. Group Activities | 5.25[e.] | 3.52 | 5.25[e.] | 4.65 | 3.52 | 4.65[e.] |
| 3. Sense of Belonging | 5.04 | 5.06 | 5.04[e.] | 4.15 | 5.06[e.] | 4.15 |

| d. Psychological Well-Being | 6.94 | 6.91 | 6.94[e] | 6.53 | 6.91[e] | 6.53 |
|---|---|---|---|---|---|---|
| 1. Fullness vs. Emptiness of Life | 7.08[e] | 6.58 | 7.08[e] | 6.43 | 6.58 | 6.43 |
| 2. Social Respect | 7.48 | 7.81 | 7.48[e] | 6.77 | 7.81[e] | 6.77 |
| 3. Personal Freedom | 6.75 | 6.76 | 6.75 | 6.93[b] | 6.76 | 6.93 |
| 4. Companionship | 7.53 | 7.48 | 7.53[e] | 7.03 | 7.48[e] | 7.03 |
| 5. Tranquility vs. Anxiety | 6.26 | 6.53[d] | 6.26[b] | 6.09 | 6.53[e] | 6.09 |
| 6. Self Approving vs. Guilty | 6.72 | 7.05[e] | 6.72[e] | 6.19 | 7.05[e] | 6.19 |
| 7. Self Confidence | 7.09[e] | 6.65 | 7.09[e] | 6.22 | 6.65[e] | 6.22 |
| 8. Elation vs. Depression | 6.65 | 6.55 | 6.65 | 6.62 | 6.55 | 6.62 |

a. two-tailed t-test for all score differences yields $p>0.100$ unless otherwise noted
b. t-test for score difference yields $p<0.100$
c. t-test for score difference yields $p<0.050$
d. t-test for score difference yields $p<0.010$
e. t-test for score difference yields $p<0.005$

Source: Compiled by authors from field data. See Chapter 3 and Appendixes E, F and G

and military preparedness are an implicit part of Israeli life whereas comparable displays would be cause for concern in the other two nations? Does the Israeli population register greater satisfaction with security because it knows that most, but not all disturbances aimed at the civilian population will be efficiently contained through collective measures? Nevertheless, the question of how a situation is perceived raises the issue of subjective versus objective indicators, and whether one is more valid than the other. The importance of either cannot be minimized, and their relationship remains an unexplored item in quality of life studies.

A second illustration of this question may be found with satisfaction with psychological well-being. The Netherlands ranks the lowest of the three nations on this dimension and most of its facets. Yet, its welfare state is the most highly developed of the three, and it is not subject to the overt sources of stress as is Israel. But, when compared to the other two nations, measures of psychological well-being indicate lower levels of satisfaction. Its housing market is the most heavily subsidized, and its citizens are highly satisfied with their public services. In addition, other cross-national studies of the quality of life have shown a positive relationship between level of economic development of a nation and the psychological well-being of its population (see Table 6.6).[24] While this finding might explain the lower scores for psychological well-being in the Netherlands versus those in the United States, it does not account for the higher scores for the Israeli sample. Perhaps the ideological cohesiveness of an embattled society is of great psychological value to Israelis despite a lower standard of living than that of the Netherlands. Other factors, such as climate, could also have influenced the responses in this survey. In fact, one Dutch market research practitioner predicted that the virtually unrelieved cloud cover present in Western Europe during the summer before the field work began would exact its gloomy toll on questions pertaining to psychological well-being in the Netherlands.[25]

Yet another example of how societies may differ in their enjoyment of various aspects of quality of life may be found in the area of satisfaction with social patterns. Sample populations in both the United States and the Netherlands registered greater levels of satisfaction with social patterns, with

TABLE 6.6 SELF-RATINGS OF HAPPINESS BY WORLD REGION: 1975

| Region | Very happy % | Fairly happy % | Not too happy % | Not ascertained % |
|---|---|---|---|---|
| North America | 40 | 51 | 8 | 1 |
| Australia | 37 | 57 | 6 | 0 |
| Western Europe | 20 | 60 | 18 | 2 |
| Latin America | 32 | 38 | 28 | 2 |
| Africa | 18 | 50 | 31 | 1 |
| Far East | 7 | 41 | 50 | 2 |

Source: George H. Gallup, 'Human Needs and Satisfactions: A Global Survey', Public Opinion Quarterly, vol.40, no.4 (1976), pp.459-67

the exception of sense of belonging, where the Israeli sample indicated a high level of satisfaction. In the Israeli context, does this one variable warrant greater weighting in the construction of the overall satisfaction score; or, once the process of national acculturation submerges this variable, what will its impact be for Israelis, who have registered far lower satisfaction scores on other measures of social patterns satisfaction?

Lastly, what can be said about the entire pattern of relationships in this survey? Despite the fact that there were differences among the sample populations, reactions to the metropolitan environments in these three nations indicate that people generally view their worlds in similar ways, an indication that it would be profitable to undertake subsequent research in other nations.

It now remains for others to determine how this exploratory study may be replicated and perhaps verified. Other important questions remain. Is too much taken for granted in the issue of study area equivalence? Have the impacts of various national policies been fully considered? Do the imperatives of culture distort responses? Can it now be stated with some confidence that the newer environments produced by metropolitan decentralization are the sine qua non of 'the good life' as perceived by representative populations in a post-industrial world? And will these changes bring about a long-term convergence of quality of life satisfaction among all the peoples of the developed nations? These questions will continue to generate a formidable agenda for future cross-national, comparative urban research.

NOTES

[1] Stefan Nowak, 'The Strategy of Cross-National Survey Research for the Development of Social Theory', in Alexander Szalai and Riccardo Petrella (eds.), Cross-National Comparative Survey Research: Theory and Practice (Pergamon Press, New York and Oxford, 1977), p.3.

[2] Alexander Szalai and Riccardo Petrella, Cross-National Comparative Survey Research, p.ix.

[3] Hadley Cantril and William Buchanan, How Nations See Each Other (University of Illinois Press, Urbana, Illinois, 1953).

[4] See Elina Almasy, Anne Balandier, and Jeanine Delatte, Comparative Survey Analysis: An Annotated Bibliography 1967-1973 (Sage Publications, Beverly Hills and London, 1976).

[5] Stefan Nowak, 'The Strategy of Cross-National Survey Research for the Development of Social Theory', p.5.

[6] Ibid., p.15; Donald N. Rothblatt, Daniel J. Garr, and Jo Sprague, The Suburban Environment and Women (Praeger, New York, 1979).

[7] See David Popenoe, 'Urban Form in Advanced Societies: A Cross-National Enquiry', in Clare Ungerson and Valerie Karn (eds.), The Consumer Experience of Housing: Cross National Perspectives (Gower, Aldershot, Hants., 1980), pp.1-20.

[8] Stefan Nowak, 'The Strategy of Cross-National Survey Research for the Development of Social Theory', p.41; Jerzy Wiatr, 'The Role of Theory in the Process of Cross-National Survey Research', in Szalai and Petrella, (eds.), Cross-National Comparative Survey Research, p.352.

[9] Stefan Nowak, 'The Strategy of Cross-National Survey Research for the Development of Social Theory', p.42.

[10] Allen D. Grimshaw, 'Comparative Sociology: In What Ways Different from Other Sociologies?' in Michael Armer and Allen D. Grimshaw (eds.), Comparative Social Research: Methodological Problems and Strategies, (John Wiley, New York, 1973), p.5.

[11] Kevin Lynch (ed.), Growing Up In Cities: Studies of the Spatial Environment of Adolescence, (MIT Press, Cambridge, Mass., and London 1977),

p.12.

[12] Charles Lewis Taylor and David A. Jodice, World Handbook of Political and Social Indicators, 3rd ed. 2 vols. (Yale University Press, New Haven and London, 1983).

[13] Ibid., I, p.89.

[14] Kent P. Schwirian, 'Some Analytical Problems in the Comparative Test of Ecological Theories', in Armer and Grimshaw, Comparative Social Research: Methodological Problems and Strategies, p.349.

[15] Ibid., p.348.

[16] William H. Lucy, 'Metropolitan Dynamics: A Cross-National Framework for Analyzing Public Policy Effects in Metropolitan Areas', Urban Affairs Quarterly, vol.11, no.2 (December, 1975), p.163.

[17] Ibid., p.161.

[18] Ibid., p.166.

[19] Cf. Peter Hall, The World Cities, 2nd ed. (McGraw Hill, New York, 1977).

[20] See, among many others, Robert C. Wood, 1400 Governments (Harvard University Press, Cambridge, Mass., 1961) and Donald N. Rothblatt, Planning the Metropolis: The Multiple Advocacy Approach (Praeger, New York, 1982).

[21] Zeline Amen Ward, 'A Policy Antimony: Public Attitudes versus Urban Conditions in Western Europe', in Michael C. Romanos (ed.), Western European Cities in Crisis (Lexington Books, Lexington, Mass., and Toronto, 1979), p.62; see also, Jacques-René Rabier, Satisfaction et Insatisfaction Quant Aux Conditions de Vie dans les Pays Membres de la Communauté Européene (Commission of the European Community, Brussels, 1974).

[22] Zeline Amen Ward, 'A Policy Antimony: Public Attitudes versus Urban Conditions in Western Europe', p.62.

[23] Frank M. Andrews, 'Comparative Studies of Life Quality: Comments on the Current State of the Art and Some Issues for Future Research', in Alexander Szalai and Frank M. Andrews (ed.), The Quality of Life: Comparative Studies (Sage Publications, Beverly Hills and London, 1980), p.275.

[24] See for example: George H. Gallup, 'Human Needs and Satisfaction: A Global Survey', Public Opinion Quarterly, vol.40, no.4 (1976), pp.459-67; Elmer Hankis, 'Structural Variables in Cross-Cultural Research on the Quality of Life', in Alexander Szalai and Frank Andrews (eds.), The Quality of Life: Comparative Studies, pp.42-56; and Alex Inkeles and Larry Diamond, 'Personal Qualities

as a Reflection of Level of National Development', paper delivered at the Ninth World Congress of the International Sociological Association, Uppsala, Sweden, August, 1978.

[25] Interview with F. D. J. Bootsma, Ogilvie Marktonderzoek, B.V., Amsterdam, October, 1978.

## APPENDIX A

## QUESTIONNAIRE SCHEDULE

Address ______________________ Date ________________

Census Tract _________________ Household Number ____

1. <u>Background Information</u>
   a. Marital Status __________

| | |
|---|---|
| Married | 5 |
| Widowed | 4 |
| Separated | 3 |
| Divorced | 2 |
| Single | 1 |

b. Duration of Marital Status (years) _________

c.1 Number of children living in your home _____
c.2 Ages __________

d. Do you rent or own this home? __________

Rent ___1___ Own ___2___

| | WOMAN | MAN (if appropriate) |
|---|---|---|
| e. Age (years) | _____ | _____ |
| f. Length of Residence in Present Home (years) | _____ | _____ |
| g. Years of Education Completed | _____ | _____ |

| | |
|---|---|
| Grade School (0-8 years) | 1 |
| High School (9-12 years) | 2 |
| Some College (13-15 years) | 3 |
| Bachelors Degree (16 years) | 4 |

| | |
|---|---|
| Some Graduate School (17 years) | 5 |
| Graduate Degree (18+ years) | 6 |

h. Occupation ______ ______

i. Please tell me the letter of the group on this card that would indicate about what the total income for your household was last year, or last month, before taxes, that is ___________

| | | | | | |
|---|---|---|---|---|---|
| a. | 0-$ 1,999 | 1 | l. | $22,000-$23,999 | 12 |
| b. | $ 2,000-$ 3,999 | 2 | m. | $24,000-$25,999 | 13 |
| c. | $ 4,000-$ 5,999 | 3 | n. | $26,000-$27,999 | 14 |
| d. | $ 6,000-$ 7,999 | 4 | o. | $28,000-$29,999 | 15 |
| e. | $ 8,000-$ 9,999 | 5 | p. | $30,000-$31,999 | 16 |
| f. | $10,000-$11,999 | 6 | q. | $32,000-$33,999 | 17 |
| g. | $12,000-$13,999 | 7 | r. | $34,000-$35,999 | 18 |
| h. | $14,000-$15,999 | 8 | s. | $36,000-$37,999 | 19 |
| i. | $16,000-$17,999 | 9 | t. | $38,000-£39,999 | 20 |
| j. | $18,000-$19,999 | 10 | u. | $40,000 and over | 21+ |
| k. | $20,000-$21,999 | 11 | | | |

j. About how far away is your place of work from here?

| | Miles | Time |
|---|---|---|
| Woman | ______ | ______ |
| Man | ______ | ______ |

k. Where did you live most of the time while you were growing up - in the country, in a small town, in a suburb of a large city, or in a large city? ___________

country __1__ small town __2__ suburb of a large city __3__ large city __4__

l.1 Was this in the San Jose area? ___________
yes __2__ no __1__

l.2 If no, where? ___________

m. About how old is the house in which you now live? ___________

n. Do you identify with any ethnic or cultural group? ___________

yes __--__ no __1__

If yes, which group? __________

How strongly do you identify with this group?

1 2 3 4 5 6 7

not strongly — very strongly

2. Housing Environment

a. Personal and Family Privacy

1. Do neighbors drop in when you'd rather they would not?

1 2 3 4 5 6 7

always — never

2. Are you bothered because of hearing your neighbors or street noises through the walls or windows?

1 2 3 4 5 6 7

always — never

3. Are you annoyed because of people looking into your windows?

1 2 3 4 5 6 7

always — never

4. Is it hard to find a place to be by yourself in the house?

1 2 3 4 5 6 7

always — never

5. Are you bothered because you don't have enough of a place where you can be out in your yard and feel that you can really have privacy from your neighbors if you want it?

1 2 3 4 5 6 7

always — never

b. Size and arrangement of the Housing Unit and Lot

| How satisfied are you with: | very unsatisfied | | | | | | very satisfied |
|---|---|---|---|---|---|---|---|
| 1. the size of your house | 1 | 2 | 3 | 4 | 5 | 6 | 7 |
| 2. the layout of your house | 1 | 2 | 3 | 4 | 5 | 6 | 7 |
| 3. the size of your lot | 1 | 2 | 3 | 4 | 5 | 6 | 7 |
| 4. the shape of your lot | 1 | 2 | 3 | 4 | 5 | 6 | 7 |
| 5. the location of your lot on your block | 1 | 2 | 3 | 4 | 5 | 6 | 7 |

6. What is the main reason for your finding your house and lot satisfying or unsatisfying?

_______________________________

7. If you could move walls around at no expense, would you be anxious to replan your house? yes __1__ no __2__

8. If yes, what would you change?

_______________________________

c. Economic and Functional Responsibilities of Home Ownership

1. How difficult do you find the financial responsibilities of owning your home?

| 1 | 2 | 3 | 4 | 5 | 6 | 7 |
|---|---|---|---|---|---|---|
| very difficult | | | | | | very easy |

2. How do you find the maintenance responsibilities for your home and lot?

| 1 | 2 | 3 | 4 | 5 | 6 | 7 |
|---|---|---|---|---|---|---|
| very difficult | | | | | | very easy |

3. What is the main reason for your finding these responsibilities easy or difficult?

   ________________________________________

d. House and Neighborhood Appearance

1. Do you think your house looks like most of the other houses within this development? yes __1__ no __2__

   How satisfied are you with this situation?

| 1 | 2 | 3 | 4 | 5 | 6 | 7 |
|---|---|---|---|---|---|---|
| very dissatisfied | | | | | | very satisfied |

2. When you go outside and look around at the streets and homes in the neighborhood, do you find what you see pleasant?

| 1 | 2 | 3 | 4 | 5 | 6 | 7 |
|---|---|---|---|---|---|---|
| very unpleasant | | | | | | very pleasant |

3. What is the main reason for your finding the appearance of your house and neighborhood pleasant or unpleasant?

   ________________________________________

e. Location

1. Would you indicate the four most important reasons (in order of importance) you have for living where you do? (Please indicate "1" next to the most important reason, and so on.)

   1) quality of public schools __________
   2) location of work __________
   3) cost of housing __________
   4) type or design of house __________
   5) social style of neighborhood __________
   6) appearance of neighborhood __________
   7) prestige of area __________
   8) close to shopping __________
   9) near friends and/or relatives __________
   10) near church or temple __________
   11) near cultural activities __________
   12) near the country __________
   13) near parks and recreation __________

2. Where did you live just before you moved to your present home - in the country, in a small town, in a suburb of a large city or in a large city?

   country __1__ small town __2__ suburb of a large city __3__ large city __4__

3. Was this in the San Jose area? yes __2__ no __1__

   If no, where? __________

4. All things considered, are you happier here, or in your previous home?

| 1 | 2 | 3 | 4 | 5 | 6 | 7 |
|---|---|---|---|---|---|---|
| much unhappier | | | about the same | | | much happier |

5. Would you indicate the four most important reasons (in order of importance) you have for feeling this way about your present home compared to your previous home? (Please indicate "1" next to the most important reason, and so on

   1) quality of public schools ________
   2) location of work ________
   3) cost of housing ________
   4) type or design of house ________
   5) social style of neighborhood ________
   6) appearance of neighborhood ________
   7) prestige of area ________
   8) close to shopping ________
   9) near friends and/or relatives ________
   10) near church or temple ________
   11) near cultural activities ________
   12) near the country ________
   13) near parks and recreation ________

3. Community Services

a. Parks and open space

1. One of the things I like about this neighborhood is that parks are no further than a convenient walk away

1 2 3 4 5 6 7

strongly disagree — no opinion — strongly agree

2. I believe nearby parks are safe

1 2 3 4 5 6 7

strongly disagree — strongly agree

3. Our parks have what I consider to be sufficient facilities and activity programs for people of all ages

1 2 3 4 5 6 7

strongly disagree — strongly agree

4. How many times during an average month do you and your family use neighborhood parks? ___________

5. On the whole, how satisfied are you with this degree of neighborhood park use?

1 2 3 4 5 6 7

very dissatisfied — very satisfied

6. What is the major reason for rating your satisfaction as you do?

______________________________________

b. Schools

1. How would you rate the academic quality of education your children are getting?

1 2 3 4 5 6 7

very low — very high

2. On the whole, how satisfied are you with the public schools that your children attend?

| 1 | 2 | 3 | 4 | 5 | 6 | 7 |
|---|---|---|---|---|---|---|
| very dissatisfied | | | | | | very satisfied |

3. What is the major reason for your satisfaction or dissatisfaction with these schools?

______________________________

c. Security/Police

1. I feel safe if I go for a walk at night in this neighborhood

| 1 | 2 | 3 | 4 | 5 | 6 | 7 |
|---|---|---|---|---|---|---|
| strongly disagree | | | | | | strongly agree |

2. I believe that my school-age children are safe when playing unsupervised in the neighborhood

| 1 | 2 | 3 | 4 | 5 | 6 | 7 |
|---|---|---|---|---|---|---|
| strongly disagree | | | | | | strongly agree |

3. On the whole, how satisfied are you with the police protection and sense of security you have in your neighborhood?

| 1 | 2 | 3 | 4 | 5 | 6 | 7 |
|---|---|---|---|---|---|---|
| very dissatisfied | | | | | | very satisfied |

4. What is the major reason for rating your satisfaction as you do?

______________________________

d. Child Care

1. Our community does enough to help the working mother

| 1 | 2 | 3 | 4 | 5 | 6 | 7 |
|---|---|---|---|---|---|---|
| very dissatisfied | | | | | | very satisfied |

2. How many times during an average week do you use day care facilities (including baby sitters)? __________

3. On the whole, how satisfied are you with day care facilities available to you?

1 2 3 4 5 6 7

very dissatisfied — very satisfied

4. What is the major reason for your satisfaction or dissatisfaction with these facilities

______________________________________

e. Transportation

1. How many cars does your family have? ____

2. How difficult is it for you to get around?

1 2 3 4 5 6 7

very difficult — very easy

3. How many times during an average week do you use public transportation? __________

4. On the whole, are you satisfied with the public transportation available for your use?

1 2 3 4 5 6 7

very dissatisfied — very satisfied

5. What is the major reason for rating your satisfaction as you do?

______________________________________

f. Entertainment and Cultural Facilities

1. There are lots of interesting places to go for an evening out

1 2 3 4 5 6 7

strongly disagree — strongly agree

2a. How many times during an average month do you go to movies, restaurants, and bars in your neighborhood? _____ b. Elsewhere in the San Jose Area? _____

3a. How many times during an average year do you go to operas, concerts, theaters, museums and art galleries in the San Jose Metropolitan Area? _____ b. Elsewhere? _____

4. On the whole, how satisfied are you with the entertainment and cultural facilities in this area?

| 1 | 2 | 3 | 4 | 5 | 6 | 7 |
|---|---|---|---|---|---|---|
| very dissatisfied | | | | | very satisfied | |

5. What is the major reason for your satisfaction or dissatisfaction with these facilities?

_____________________________________

6. What other kind of entertainment activity do you and/or your family participate in?

_____________________________________

g. Finance of Services

1. Of all these services, which would you be willing to spend more money on (either tax dollars or your own)? Please check the appropriate services listed below.
   a) Parks and Open Space _______________
   b) Schools _______________
   c) Security/Police _______________
   d) Child Care _______________
   e) Transportation _______________
   f) Entertainment and Cultural Facilities _______________
   g) Other (specify) _______________

4. Social Patterns

a. Friendships

1. How many people within this neighborhood (within 5 minutes from your home) do you speak to? __________

2. With how many of your immediate neighbors could you:

   a. talk to about topics of general interest? ____________
   b. borrow things from? ____________
   c. call on for aid during a crisis (such as illness), child care, errands, etc.? ____________
   d. ask to help with a project such as home repair? ____________
   e. talk to about a personal problem? ____________

3. How many close friends do you have:

   a. in your neighborhood? ____________
   b. in other sections of the San Jose Metropolitan Area (Santa Clara County)? ____________
   c. outside the San Jose Metropolitan Area, but within an hour's drive? ____________
   d. farther away? ____________

4. How difficult do you find meeting people and establishing friendships?

| 1 | 2 | 3 | 4 | 5 | 6 | 7 |
|---|---|---|---|---|---|---|
| very difficult | | | | | | very easy |

5. What is the major reason for your finding meeting people and establishing friendships easy or difficult?

____________________________________________

6. On the whole, how satisfied are you with:

   a. the kind of friendships you have in your neighborhood?

| very dissatisfied | | | | | | very satisfied |
|---|---|---|---|---|---|---|
| 1 | 2 | 3 | 4 | 5 | 6 | 7 |

b. the kind of friendships you have in the rest of the San Jose Metropolitan Area?

| | very dissatisfied | | | | | very satisfied |
|---|---|---|---|---|---|---|
| 1 | 2 | 3 | 4 | 5 | 6 | 7 |

7. What is the major reason for your satisfaction or dissatisfaction with these patterns of friendships?

_______________________________________

b. Group Participation

1. How involved are you in each of these kinds of groups?

| | not involved at all | | | | | | organizer or leader |
|---|---|---|---|---|---|---|---|
| a. informal social groups in your neighborhood (bridge clubs, etc.) | 1 | 2 | 3 | 4 | 5 | 6 | 7 |
| b. action projects in your neighborhood (co-ops, child care, improvement assoc.) | 1 | 2 | 3 | 4 | 5 | 6 | 7 |
| c. groups related to your children's activities (P.T.A., cub scouts) | 1 | 2 | 3 | 4 | 5 | 6 | 7 |
| d. religious groups | 1 | 2 | 3 | 4 | 5 | 6 | 7 |
| e. political organizations | 1 | 2 | 3 | 4 | 5 | 6 | 7 |
| f. informal groups outside your neighborhood | 1 | 2 | 3 | 4 | 5 | 6 | 7 |
| g. action groups outside your neighborhood (League of Women Voters, Sierra Club) | 1 | 2 | 3 | 4 | 5 | 6 | 7 |

h. organized social groups outside your neighborhood (Sororities, lodges, clubs) 1 2 3 4 5 6 7

i. other organized groups 1 2 3 4 5 6 7

2. On the whole, how satisfied are you with:

very dissatisfied — very satisfied

a. the kind of opportunities for group participation in your neighborhood? 1 2 3 4 5 6 7

b. the kind of opportunities for group participation available in the rest of the San Jose Metropolitan Area? 1 2 3 4 5 6 7

3. What is the major reason for your satisfaction or dissatisfaction with these group involvements?

________________________________________

c. Sense of Belonging

1. How closely do you identify with (feel in tune with) your neighborhood?

1 2 3 4 5 6 7

not closely at all — very closely

2. On the whole, how satisfied are you with this degree of identification?

1 2 3 4 5 6 7

very dissatisfied — very satisfied

3. What is the major reason for rating your satisfaction as you do?

   ______________________________

5. Psychological Well-Being
   Please indicate your personal feelings (ranked 1 to 10) for the following categories:

   a. Fullness vs. Emptiness of Life (how emotionally satisfying, abundant or empty, your life felt during the past month).
      10. Consummate fulfillment and abundance.
      9. Replete with life's abundant goodness.
      8. Filled with warm feelings of contentment and satisfaction.
      7. My life is ample and satisfying.
      6. Life seems fairly adequate and relatively satisfying.
      5. Some slight sense of lack, vague and midly troubling.
      4. My life seems deficient, dissatisfying.
      3. Life is pretty empty and barren.
      2. Desolate, drained dry, impoverished.
      1. Gnawing sense of emptiness, hollowness, void.

   b. Social Respect vs. Social Contempt (how you felt other people regarded you, or felt about you, during the past month).
      10. Excite the admiration and awe of everyone who matters.
      9. Stand extremely high in the estimation of people whose opinions count with me.
      8. People I admire recognize and respect my good points.
      7. Confident that some people think well of me.
      6. Feel I am appreciated and respected to some degree.
      5. Some people don't seem to see much value in me.
      4. I am looked upon as being of small or of no account.
      3. People have no respect for me at all.
      2. I am scorned, slighted, pushed aside.
      1. Everyone despises me and holds me in contempt.

   c. Personal Freedom vs. External Constraint (how much you felt you were free or not free to do as you wanted).

10. Absolutely free to consider and try any new and adventuresome prospect.
9. Independent and free to do as I like.
8. Ample scope to go my own way.
7. Free, within broad limits, to act much as I want to.
6. Can do a good deal on my own initiative and in my own fashion. No particularly restrictive limitations.
5. Somewhat constrained and hampered. Not free to do things my own way.
4. Checked and hindered by too many demands and constraints.
3. Hemmed in, cooped up. Forced to do things I don't want to do.
2. Trapped, oppressed.
1. Overwhelmed, smothered. Can't draw a free breath.

d. Companionship vs. Being Isolated (the extent to which you felt emotionally accepted by, or isolated from, other people).
10. Complete participation in warm, intimate friendship.
9. Enjoy the warmth of close companionship.
8. Thoroughly and genuinely liked.
7. Feel accepted and liked.
6. More or less accepted.
5. Feel a little bit left out.
4. Feel somewhat neglected and lonely.
3. Very lonely. No one seems to care much about me.
2. Tremendously lonely. Friendless and forlorn.
1. Completely isolated and forsaken. Abandoned. Ache with loneliness.

e. Tranquility vs. Anxiety (how calm or troubled you felt).
10. Perfect and complete tranquility. Unshakably secure.
9. Exceptional calm, wonderfully secure and carefree.
8. Great sense of well-being, Essentially secure, and very much at ease.
7. Pretty generally secure and free from care.
6. Nothing particularly troubling me: more or less at ease.
5. Somewhat concerned with minor worries or problems. Slightly ill-at-ease a bit

troubled.
4. Experiencing some worry, fear, trouble, or uncertainty. Nervous, jittery, on edge.
3. Considerable insecurity. Very troubled by significant worries, fears, uncertainties.
2. Tremendous anxiety and concern. Harassed by major worries and fears.
1. Completely beside myself with dread, worry, fear. Overwhelmingly distraught and apprehensive. Obsessed or terrified by insoluble problems and fears.

f. Personal Moral Judgment (how self-approving, or how guilty, you felt).
10. Have a transcendent feeling of moral perfection and virtue.
9. I have a sense of extraordinary worth and goodness.
8. In high favor with myself. Well up to my own best standards.
7. Consider myself pretty close to my own best self.
6. By and large, measuring up to most of my moral standards.
5. Somewhat short of what I ought to be.
4. I have a sense of having done wrong.
3. Feel that I have failed morally.
2. Heavy laden with my own moral worthlessness.
1. In anguish. Tormented by guilt and self-loathing.

g. Self-Confidence vs. Feeling of Inadequacy (how self-assured and adequate, or helpless and inadequate, you felt).
10. Nothing is impossible to me. Can do anything I want.
9. Feel remarkable self-assurance. Sure of my superior powers.
8. Highly confident of my capabilities.
7. Feel my abilities sufficient and my prospects good.
6. Feel fairly adequate.
5. Feel my performance and capabilities somewhat limited.
4. Feel rather inadequate.
3. Distressed by my weakness and lack of ability.
2. Wretched and miserable. Sick of my own

incompetence.
1. Crushing sense of weakness and futility. I can do nothing.

h. Elation vs. Depression (how elated or depressed, happy or unhappy, you felt during the past month).
10. Complete elation. Rapturous joy and soaring ecstasy.
9. Very elated and in very high spirits. Tremendous delight and buoyancy.
8. Elated and in high spirits.
7. Feeling very good and cheerful.
6. Feeling pretty good, "O.K."
5. Feeling a little bit low. Just so-so.
4. Spirits low and somewhat "blue".
3. Depressed and feeling very low. Definitely "blue".
2. Tremendously depressed. Feeling terrible, miserable "just awful".
1. Utter depression and gloom. Completely down. All is black and leaden.

## APPENDIX B

## DEPENDENT VARIABLES

a. Housing Environment
   1. Personal and Family Privacy
      a. Unwanted Visits
      b. Audio Privacy
      c. Visual Privacy
      d. Indoor Privacy
      e. Outdoor Privacy

   2. Size and Arrangement of Housing Unit and Lot
      a. Size of House
      b. Layout of House
      c. Size of Lot
      d. Shape of Lot
      e. Location on Block

   3. Economic and Functional Responsibilities of Home Ownership
      a. Financial Responsibilities
      b. Maintenance Responsibilities

   4. House and Neighborhood as a Physical Symbol of Tastes and Personality
      a. House Alikeness
      b. Neighborhood Appearance

b. Community Services
   1. Parks and Open Space
      a. Accessibility
      b. Safety
      c. Accommodates all Age Groups
      d. Satisfaction with Usage

   2. Schools
      a. Academic Quality
      b. Satisfaction with Schools

3. Security/Police
   a. Safe Walking at Night
   b. Children Safe in Neighborhood
   c. Satisfaction with Security

4. Child Care
   a. Community Helps Working Mother
   b. Satisfaction with Day Care Facilities

5. Transportation
   a. Ease of Transport
   b. Satisfaction with Transit

6. Entertainment and Culture
   a. Availability of Entertainment and Cultural Facilities
   b. Satisfaction with Entertainment and Cultural Facilities

c. <u>Social Patterns</u>
1. Friendships
   a. Difficulty Making Friends
   b. Satisfaction with Friends in Neighborhood
   c. Satisfaction with Friends in Metropolitan Area

2. Group Activity
   a. Satisfaction with Group Opportunities in Neighborhood
   b. Satisfaction with Group Opportunities in Metropolitan Area

3. Sense of Belonging
   a. Identification with Neighborhood
   b. Satisfaction with Neighborhood Identification

d. <u>Psychological Well-Being</u>

1. Fullness vs. Emptiness of Life
2. Receptivity towards the world
3. Social Respect vs. Social Contempt
4. Personal Freedom vs. Constraint
5. Sociability vs. Withdrawal
6. Companionship vs. Isolation
7. Work Satisfaction
8. Tranquility vs. Anxiety
9. Self Approving vs. Guilty
10. Self Confidence vs. Inadequacy
11. Energy vs. Fatigue
12. Elation vs. Depression

# APPENDIX C

## INDEPENDENT VARIABLES

a. <u>Social Class Influences</u>
   1. Income
   2. Education
   3. Occupational Status

b. <u>Subcultural Influences</u>
   1. Ethnic Identity
   2. Regional Origin
      a. Extent of Urban Origins
      b. Distance to San Jose Area
   3. Length of Residency

c. <u>Life Cycle Influences</u>
   1. Age
   2. Marital Status
   3. Number and Age of Children

d. <u>Suburban Environmental Influences</u>
   1. Residential Density
   2. Household Distance from the Central City Center
   3. Age of Neighborhood
   4. Distance to Work
   5. Mean Neighborhood Income and Educational Levels
      a. Ratio of Family Income to Mean Neighborhood Family Income
      b. Ratio of Woman's Education to Mean Neighborhood Woman's Education
   6. Population Size of Political Unit
   7. Design and Site Plan Characteristics

APPENDIX D

EXPLORATORY VARIABLES

a. Housing Environment
   1. Desire to Replan House
   2. Reasons for Present Location
      a. First
      b. Second
      c. Third
      d. Fourth
   3. Extent of Urban Character of Prior Location
   4. Distance to San Jose Area of Prior Location
   5. Happier at Present Location
   6. Reasons for Satisfaction of Present Home
      a. First
      b. Second
      c. Third
      d. Fourth

b. Community Services
   1. Park Usage per Month
   2. Child Care Usage per Week
   3. Number of Cars
   4. Public Transport Usage per Week
   5. Evenings Out per Month in Neighborhood
   6. Evenings Out per Month in Metropolitan Area
   7. Cultural Events per Year in Metropolitan Area
   8. Cultural Events per Year Elsewhere
   9. Willingness to Spend More Money on:
      a. Parks and Open Space
      b. Schools
      c. Security Police
      d. Child Care
      e. Transportation
      f. Entertainment and Cultural Facilities
      g. Other

c. Social Patterns
1. Speaking Acquaintances in Neighborhood
2. Neighbors to Talk about General Topics
3. Neighbors to Borrow from
4. Neighbors to Aid in Crises
5. Neighbors to Help in Project
6. Neighbors to Discuss Personal Problems
7. Close Friends in Neighborhood
8. Close Friends in Metropolitan Area
9. Close Friends 1 Hour away
10. Close Friends Farther away
11. Involved in Neighborhood Informal Groups
12. Involved in Neighborhood Action Groups
13. Involved in Children's Activity Groups
14. Involved in Religious Groups
15. Involved in Political Organizations
16. Involved in Outside Informal Groups
17. Involved in Outside Action Groups
18. Involved in Outside Social Organization
19. Involved in other Organized Groups

d. Background
1. Duration of Marital Status
2. Rent or Own
3. Age of Man
4. Length of Residency Man
5. Education Man
6. Working Woman (% Working)
7. Occupation Status of Man
8. Ethnic Identity (% Identified)
9. Man's Education to Mean Neighborhood Man's Education
10. Woman's Occupation to Mean Neighborhood Woman's Occupation
11. Man's Occupation to Mean Neighborhood Man's Occupation
12. Occupation-Education Ratio Woman
13. Occupation-Education Ratio Man
14. Education Ratio Woman-Man
15. Occupation Ratio Woman-Man

APPENDIX E

DEPENDENT VARIABLE MEAN SCORES: U.S.A., ISRAEL AND THE NETHERLANDS

| Dependent Variable | SCORE[a] | | |
|---|---|---|---|
| | U.S.A. (n=825) | Israel (n=295) | Netherlands (n=215) |
| HOUSING ENVIRONMENT | 5.59 | 4.80 | 5.12 |
| 1. PRIVACY | 5.82 | 5.39 | 4.92 |
| a. unwelcomed visits | 5.98 | 5.71 | 6.30 |
| b. audio privacy | 5.55 | 4.99 | 5.39 |
| c. visual privacy | 6.42 | 6.13 | 5.56 |
| d. indoor privacy | 5.65 | 5.73 | 3.38 |
| e. outdoor privacy | 5.50 | 4.39 | 3.99 |
| 2. HOUSE AND LOT | 5.56 | 5.09 | 5.49 |
| a. size of house | 5.51 | 4.81 | 5.10 |
| b. layout of house | 5.63 | 5.18 | 5.41 |
| c. size of lot | 5.15 | 5.08 | 5.11 |
| d. shape of lot | 5.57 | 5.01 | 5.23 |
| e. location on block | 5.96 | 5.30 | 5.93 |
| 3. HOMEOWNER RESPONSIBILITIES | 5.22 | 4.20 | 4.87 |
| a. financial responsibilities | 5.13 | 4.18 | 5.15 |
| b. maintenance responsibilities | 5.31 | 4.20 | 4.61 |
| 4. SYMBOL OF TASTE | 5.77 | 4.57 | 5.17 |
| a. house alikeness | 5.62 | 4.43 | 5.42 |
| b. neighborhood appearance | 5.91 | 4.70 | 4.92 |
| COMMUNITY SERVICES | 5.10 | 5.12 | 5.12 |
| 1. PARKS AND OPEN SPACES | 4.97 | 5.34 | 5.34 |
| a. accessibility | 5.08 | 4.82 | 5.76 |
| b. safety | 5.00 | 5.17 | 4.83 |
| c. accommodate all ages | 4.71 | 3.03 | 5.42 |
| d. satisfaction with usage | 5.02 | 4.36 | 5.35 |

Appendix E (continued)

| Dependent Variable | SCORE[a.] | | |
|---|---|---|---|
| | U.S.A. (n=825) | Israel (n=295) | Netherlands (n=215) |
| 2. SCHOOLS | 5.18 | 4.83 | 5.79 |
| a. academic quality | 5.25 | 4.88 | 5.73 |
| b. satisfaction with schools | 5.10 | 4.79 | 5.85 |
| 3. SECURITY/POLICE | 5.24 | 5.14 | 4.32 |
| a. safe walking at night | 5.08 | 5.10 | 4.17 |
| b. children safe in neighborhood | 5.38 | 5.22 | 4.46 |
| c. satisfaction with security | 5.22 | 5.07 | 4.29 |
| 4. CHILDCARE | 4.18 | 3.77 | 4.24 |
| a. community helps working mother | 4.15 | 3.28 | 4.27 |
| b. satisfaction with daycare facilities | 4.26 | 4.27 | 4.24 |
| 5. TRANSPORTATION | 5.42 | 3.70 | 6.01 |
| a. ease of transit | 6.46 | 4.07 | 6.10 |
| b. satisfaction with transit | 4.37 | 3.45 | 5.93 |
| 6. ENTERTAINMENT AND CULTURE | 5.49 | 2.93 | 5.02 |
| a. availability | 5.55 | 3.09 | 5.05 |
| b. satisfaction | 5.42 | 2.71 | 4.98 |
| SOCIAL PATTERNS | 5.33 | 4.42 | 4.71 |
| 1. FRIENDSHIPS | 5.66 | 4.78 | 5.25 |
| a. difficulty making friends | 5.31 | 4.42 | 4.33 |
| b. satisfaction neighborhood friends | 5.64 | 4.83 | 5.41 |
| c. satisfaction metropolitan area friends | 6.02 | 5.11 | 5.95 |
| 2. GROUP PARTICIPATION | 5.25 | 3.52 | 4.65 |
| a. satisfaction group opportunities neighbor-hood | 5.09 | 3.69 | 4.69 |
| b. satisfaction group opportunities metro-politan area | 5.42 | 3.41 | 4.65 |
| 3. SENSE OF BELONGING | 5.04 | 5.07 | 4.14 |
| a. identification with neighborhood | 4.65 | 5.00 | 3.55 |
| b. satisfaction with neighborhood identity | 5.42 | 5.13 | 4.73 |

| Dependent Variable | SCORE[a] | | |
|---|---|---|---|
| | U.S.A. (n=825) | Israel (n=295) | Netherlands (n=215) |
| PSYCHOLOGICAL WELL-BEING | 6.94 | 6.91 | 6.53 |
| 1. FULLNESS vs EMPTINESS | 7.08 | 6.58 | 6.43 |
| 2. SOCIAL RESPECT vs CONTEMPT | 7.48 | 7.81 | 6.77 |
| 3. PERSONAL FREEDOM vs CONSTRAINT | 6.75 | 6.76 | 6.93 |
| 4. COMPANIONSHIP vs ISOLATION | 7.53 | 7.48 | 7.03 |
| 5. TRANQUILITY vs ANXIETY | 6.26 | 6.53 | 6.09 |
| 6. SELF APPROVING vs GUILTY | 6.72 | 7.05 | 6.19 |
| 7. SELF CONFIDENCE vs INADEQUACY | 7.09 | 6.65 | 6.22 |
| 8. ELATION vs DEPRESSION | 6.65 | 6.55 | 6.62 |

a. Range of possible scores are 1.00 to 7.00 for all variables except for Physological Well-Being variables which cover a range of 1.00 to 10.00

Source: Compiled by the authors from field data.

## APPENDIX F

## INDEPENDENT VARIABLE MEAN SCORES: U.S.A., ISRAEL AND THE NETHERLANDS

| Independent Variable | SCORE | | |
|---|---|---|---|
| | U.S.A. (n=825) | Israel (n=295) | Netherlands (n=215) |
| SOCIAL CLASS INFLUENCES | | | |
| 1. Income | 24,400($)[a] | 98,100(I£)[b] | 27,100(HFL)[c] |
| 2. Education[d] | 3.25 | 2.71 | 1.69 |
| 3. Occupational Status[e] | 2.56 | 3.12 | 1.48 |
| SUBCULTURAL INFLUENCES | | | |
| 1. Ethnic Identity (1.0= Lowest Identity; 7.0=Highest) | 1.80 | 2.41 | 2.31 |
| 2. Regional Origin | | | |
| a. Extent of Urban Origins[f] | 2.70 | 3.01 | 3.45 |
| b. Distance to Metropolitan Area[g] | 4.09 | 2.41 | 1.38 |
| 3. Length of Residency (years) | 4.54 | 4.23 | 7.37 |
| LIFE CYCLE INFLUENCES | | | |
| 1. Age (years) | 34.70 | 30.52 | 37.60 |
| 2. Marital Status (% married) | 85.4 | 96.4 | 91.6 |
| 3. Children | | | |
| a. Number | 2.12 | 2.32 | 2.05 |
| b. Age | 8.55 | 5.39 | 9.95 |
| SUBURBAN ENVIRONMENTAL INFLUENCES | | | |
| 1. Residential Density | 6-20 | 24-35 | 30-37 |
| 2. Distance from Central City | 2-8 | 2-9 | 2-4 |
| 3. Age of Neighborhood | 13.36 | 5.96 | 19.59 |

| Independent Variable | SCORE | | |
|---|---|---|---|
| | U.S.A. (n=825) | Israel (n=295) | Netherlands (n=215) |
| 4. Travel Time to Work (minutes) | | | |
| a. Woman's | 8.77 | 17.24 | 4.59 |
| b. Man's | 21.93 | 32.89 | 24.07 |
| 5. Population Size of Political Unit (thousands) | 22-552 | 25-343 | 53-458 |
| 6. Design and Site Plan Characteristics (P= Planned; LP=Less Planned) | P & LP | P & LP | P & LP |

a. 1976 income in U.S. Dollars reported during field interviews held in 1977

b. 1978 income in Israel Pounds reported during field interviews held in 1978 ($1=16I£)

c. 1978 income in Dutch Guilders reported during field interviews held in 1978 ($1= 2HFL)

d. For years of education completed the following scores are applicable:

1.0 = Grade School (0 - 8 years)
2.0 = High School (9 - 12 years)
3.0 = Some College (13 - 15 years)
4.0 = College Graduate (16 years)
5.0 = Some Graduate School (17 years)
6.0 = Graduate Degree (18 years)

e. For occupational status, scores were estimated by research staff from descriptions of employment activities on questionnaires using the following scale:

1.0 = Housewife
2.0 = Unskilled Worker
3.0 = Semi-Skilled Worker
4.0 = Skilled Worker
5.0 = Manager or Owner
6.0 = Professional

f. Extent of urban origins is represented with the following scores:

1.0 = country
2.0 = small town
3.0 = suburb of large city
4.0 = large city

g. Distance of regional origins to the area of residence is based on actual distance measurement and the following scale:

Appendix F (continued)

<u>In the United States</u>

1.0 = within the San Jose metropolitan area (Santa Clara County)
2.0 = within a 50 mile radius
3.0 = 51 - 250 miles
4.0 = 251 - 500 miles
5.0 = 501 - 1,000 miles
6.0 = 1,001 - 1,500 miles
7.0 = 1,501 - 2,000 miles
8.0 = 2,001 - 3,000 miles
9.0 = 3,001 + miles

<u>In Israel and the Netherlands</u>

1.0 = within the metropolitan area of residence
2.0 = within a 100 kilometer radius
3.0 = 101 - 300 kilometers
4.0 = 301 + kilometers
5.0 = other country beyond 500 kilometers one can easily visit
6.0 = other country one can not easily visit

## APPENDIX G

## EXPLORATORY VARIABLE MEAN SCORES: U.S.A., ISRAEL AND THE NETHERLANDS

| Exploratory Variable | SCORE | | |
|---|---|---|---|
| | U.S.A. (n=825) | Israel (n=295) | Netherlands (n=215) |
| a. Housing Environment | | | |
| 1. Desire to Replan House (% no.) | 55 | 49 | 57 |
| 2. Reasons for Present Location[a]. | | | |
| a. First | h.c. | h.d. | p.n. |
| b. Second | h.c. | s.n. | a.n. |
| c. Third | a.n. | a.n. | p.n. |
| d. Fourth | a.n. | p.n. | shop |
| 3. Extent of Urban Character of Prior Location[b]. | 3.00 | 3.13 | 3.65 |
| 4. Distance to Metro Area of Prior Location[c]. | 1.98 | 1.65 | 1.18 |
| 5. Happier at Present Location | 5.79 | 5.20 | 5.46 |
| 6. Reasons for Satisfaction of Present Home[a]. | | | |
| a. First | h.d. | h.d. | h.d. |
| b. Second | h.d. | h.c. | h.d. |
| c. Third | a.n. | work | h.c. |
| d. Fourth | a.n. | work | h.c. |
| b. Community Services | | | |
| 1. Park Usage per Month | 4.56 | 7.59 | 5.44 |
| 2. Child Care Usage per Week | 1.47 | 2.06 | 0.25 |
| 3. Number of Cars | 1.91 | 0.77 | 0.94 |
| 4. Public Transport Usage | | | |

Appendix G (continued)

| Exploratory Variable | SCORE | | |
|---|---|---|---|
| | U.S.A. (n=825) | Israel (n=295) | Netherlands (n=215) |
| per Week | 0.40 | 5.12 | 1.58 |
| 5. Evenings Out per Month Neighborhood | 2.84 | 0.98 | 0.57 |
| 6. Evenings Out Per Month Metro Area | 2.73 | 2.16 | 0.79 |
| 7. Cultural Events per Year Metro Area | 5.19 | 6.19 | 2.33 |
| 8. Cultural Events per Year Elsewhere | 3.02 | 0.96 | 0.78 |
| c. Social Patterns | | | |
| 1. Speaking Acquaintanceships | 22.67 | 4.57 | 3.19 |
| 2. Neighbors and General Interest | 9.06 | 4.45 | 3.16 |
| 3. Neighbors to borrow from | 4.30 | 3.18 | 3.28 |
| 4. Neighbors to aid in crises | 4.67 | 3.10 | 3.36 |
| 5. Neighbors to help in project | 2.59 | 2.05 | 1.46 |
| 6. Neighbors to discuss personal problems | 1.58 | 1.01 | 0.78 |
| 7. Close friends in Neighborhood | 2.33 | 5.41 | 2.19 |
| 8. Close friends in Metro. Area | 8.72 | 7.10 | 4.20 |
| 9. Close friends 1 hour away | 5.20 | 4.46 | 1.89 |
| 10. Close friends farther away | 8.23 | 3.56 | 2.05 |
| 11. Involved in Neighborhood Informal Groups[d.] | 2.09 | 2.43 | 1.53 |
| 12. Involved in Neighborhood Action Groups[d.] | 2.18 | 2.63 | 1.29 |
| 13. Involved in Children's Activity Groups[d.] | 3.52 | 2.63 | 2.19 |
| 14. Involved in Religious Groups[d.] | 2.64 | 1.16 | 1.78 |
| 15. Involved in Political Orgs.[d.] | 1.67 | 1.31 | 1.29 |
| 16. Involved in Outside Informal Groups[d.] | 2.80 | 2.31 | 1.48 |
| 17. Involved in Outside Action Groups[d.] | 1.86 | 1.20 | 1.06 |

| Exploratory Variable | SCORE | | |
|---|---|---|---|
| | U.S.A. (n=825) | Israel (n=295) | Netherlands (n=215) |
| 18. Involved in Outside Social Org.[d.] | 1.95 | 1.36 | 1.12 |
| 19. Involved in Other Org. Groups[d.] | 2.02 | 1.36 | 1.67 |
| d. Background | | | |
| 1. Duration of Marital Status (years) | 11.04 | 8.49 | 13.95 |
| 2. Rent or Own (% Own) | 91 | 82 | 16 |
| 3. Age of Man | 37.30 | 33.39 | 40.63 |
| 4. Length of Residency of Man | 4.73 | 4.34 | 7.28 |
| 5. Education of Man[e.] | 3.95 | 3.17 | 1.96 |
| 6. Working Women (% working) | 51.3 | | |
| 7. Occupation Status of Man[f.] | 4.71 | 4.31 | 3.66 |
| 8. Ethnic Identity | 1.83 | 2.41 | 2.31 |
| 9. Occupation-Education Ratio of Woman | 0.81 | 1.18 | 1.01 |
| 10. Occupation-Education Ratio of Man | 1.29 | 1.57 | 2.22 |
| 11. Education Ratio Woman-Man | 0.87 | 0.93 | 0.93 |
| 12. Occupation Ratio Woman-Man | 0.54 | 0.72 | 0.43 |

a. Legend:
sch. = quality of public schools
work = location of work
h.c. = cost of housing
h.d. = type or design of house
s.n. = social style of neighborhood
a.n. = appearance of neighborhood
p.n. = prestige of neighborhood
shop = close to shopping
friend = near friends or relatives
park = near parks and recreation

b. Extent of urban origins is represented with the following scores:
1.0 = country
2.0 = small town
3.0 = suburb of large city
4.0 = large city

c. Distance of regional origins to the area of residence is

based on actual distance measurement and the following scale:

<u>In the United States</u>

1.0 = within the San Jose Metropolitan Area (Santa Clara County)
2.0 = within a 50 mile radius
3.0 = 51 - 250 miles
4.0 = 251 - 500 miles
5.0 = 501 - 1,000 miles
6.0 = 1,001 - 1,500 miles
7.0 = 1,501 - 2,000 miles
8.0 = 2,001 - 3,000 miles
9.0 = 3,001 + miles

<u>In Israel and the Netherlands</u>

1.0 = within the metropolitan area of residence
2.0 = within a 100 kilometer radius
3.0 = 101 - 300 kilometers
4.0 = 301+ kilometers
5.0 = other country beyond 500 kilometers one can easily visit
6.0 = other country one cannot easily visit

d. Range of possible scores are 1.00 for no involvement in group activities to 7.00 for organizing or leadership role.

e. For years of education completed, the following scores are applicable:

1.0 = Grade School (0 - 8 years)
2.0 = High School (9 - 12 years)
3.0 = Some College (13 - 15 years)
4.0 = College Graduate (16 years)
5.0 = Some Graduate School (17 years)
6.0 = Graduate Degree (18 years)

f. For occupational status, scores were estimated by research staff from descriptions of employment activities on questionnaires using the following scale:

1.0 = Housewife
2.0 = Unskilled Worker
3.0 = Semi-skilled Worker
4.0 = Skilled Worker
5.0 = Manager or Owner
6.0 = Professional

# BIBLIOGRAPHY

## BOOKS

Abert, J. G., Economic Policy and Planning in the Netherlands, 1950-1965 (Yale University Press, New Haven, 1969)

Abrams, C., The Future of Housing (Harper and Brothers, New York, 1946)

Allan, G. A., A Sociology of Friendship and Kinship (Allen and Unwin, London, 1979)

Almasy, E., Balandier, A. and Delatte, J., Comparative Survey Analysis: An Annotated Bibliography 1967-1973 (Sage, Beverly Hills and London, 1976)

Aronoff, M. J., Frontiertown: The Politics of Community Building in Israel (Manchester University Press, Manchester, 1974)

Berler, A. et al., Urban-Rural Relations in Israel: Social and Economic Aspects (Settlement Study Center, Rehovot, 1970)

--- New Towns in Israel (Israel Universities Press, Jerusalem, 1970)

Berry, B. J. L., Urbanization and Counterurbanization (Sage, Beverly Hills, 1976)

Bish, R. L., The Public Economy of Metropolitan Areas (Markham, Chicago, 1971)

--- and Nourse, H. O., Urban Economics and Policy Analysis (McGraw Hill, New York, 1975)

Bollens, J. C. and Schmandt, H. J., The Metropolis: Its People, Politics and Economic Life (Harper and Row, New York, 1975, 4th ed., 1982)

Burke, G. L., Greenheart Metropolis: Planning the Western Netherlands (St. Martin's Press, New York, 1966)

Butler, R. N. and Lewis, M. I., Aging and Mental Health (C. V. Mosby, St. Louis, 1977)

Campbell, A., Converse, P. E., and Rodgers, W. L.,

The Quality of American Life (Russell Sage Foundation, New York, 1976)
Cantril, H. and Buchanan, W., How Nations See Each Other (University of Illinois Press, Urbana, 1953)
--- The Pattern of Human Concerns (Rutgers University Press, New Brunswick, 1965)
Clark, S.D., The Suburban Society (University of Toronto Press, Toronto, 1966)
Clout, H. D. (ed.), Regional Development in Western Europe (John Wiley, New York, 1981)
Creese, W. L., The Search for Environment: The Garden City - Before and After (Yale University Press, New Haven, 1966)
Darin-Drabkin, H., Housing in Israel: Economic and Sociological Aspects (Gadish Books, Tel-Aviv, 1957)
--- (ed.), Public Housing in Israel: Surveys and Evaluations of Activities in Israel's First Decade (1948-1958) (Gadish Books, Tel-Aviv, 1959)
--- Land Policy and Urban Growth (Pergamon Press, Oxford and New York, 1977)
Dash, J. and Efrat, E., The Israel Physical Master Plan (Israel Ministry of the Interior, Jerusalem, 1964)
Economic and Social Opportunities, Inc., Female Heads of Household and Poverty in Santa Clara County (Economic and Social Opportunities, San Jose, 1974)
Efrat, E., Urbanization in Israel (Croom Helm, London, 1984)
Eichler, E. P. and Kaplan, M., The Community Builders (University of California Press, Berkeley, 1967)
Epstein, C., Woman's Place (University of California Press, Berkeley, 1970)
Ewald, W. R., Jr., Environment for Man (Indiana University Press, Bloomington, 1967)
Fischer, C. S., To Dwell Among Friends: Personal Networks in Town and City (The University of Chicago Press, Chicago, 1982)
Gakenheimer, R. (ed.), The Automobile and the Environment: An International Perspective (MIT Press, Cambridge, Mass., 1978)
Gallanty, E. Y., New Towns, Antiquity to the Present (George Braziller, New York, 1975)
Gallion, A. B. and Eisner, S., The Urban Pattern (D. Van Nostrand Company, New York, 1975)
Galnoor, I., Steering the Policy: Communications and Politics in Israel (Sage, Beverly Hills and

London, 1982)
Gans, H. J., The Urban Villagers (Free Press of Glencoe, New York, 1962)
--- The Levittowners (Vintage Books, New York, 1967)
Glazer, N. and Moynihan, D. P., Beyond the Melting Pot (MIT Press, Cambridge, Mass., 1963)
Greer, S., The Urbane View: Life and Politics in Metropolitan America (Oxford University Press, New York, 1972)
Griffiths, R. T. (ed.), The Economy and Politics of the Netherland since 1945 (Martins Nijhoff, The Hague, 1980)
Grinberg, D. I. Housing in the Netherlands 1900-1940 (Delft University Press, Delft, 1977)
Haar, C. (ed.), The President's Task Force on Suburban Problems (Ballinger Publishing Company, Cambridge, Mass., 1974)
Halevi, N. and Klinov-Malul, R., Economic Development of Israel (Frederick A. Praeger, New York, 1968)
Hall, P. et al., The Containment of Urban England (2 volumes, George Allen and Unwin, London, 1973)
--- and Hay, D., Growth Centers in the European Urban System (University of California Press, Berkeley, 1980)
--- The World Cities, 2nd edition (McGraw Hill, New York, 1977)
Hapgood, K. and Getzels, J. (eds.), Planning, Women and Change (American Society of Planning Officials, Chicago, 1974)
Hayden, D., Redesigning the American Dream: The Future of Housing, Work, and Family Life (W. S. Norton, New York, 1984)
Hazelton, L., Israeli Women: The Reality Behind the Myths (Simon and Schuster, New York, 1977)
Heilbrun, J., Urban Economics and Public Policy, 2nd ed. (St. Martin's Press, New York, 1981)
Hirschman, A. O., The Strategy of Economic Development (Yale University Press, New Haven, 1958)
Hoover, E. M. and Vernon, R., Anatomy of a Metropolis (Harvard University Press, Cambridge, Mass., 1959)
Johnson, E. A. J., The Organization of Space in Developing Countries (Harvard University Press, Cambridge, Mass., 1970)
Kreps, J., Sex in the Market Place (Johns Hopkins Press, Baltimore, 1971)
LaGory, M. and Pipkin, J., Urban Social Space (Wadsworth Publishing, Belmont, Cal., 1981)

Lapin, J. S., Structuring the Journey to Work (University of Pennsylvania Press, Philadelphia, 1964)
Lampl, P., Cities and Planning in the Ancient Near East (George Braziller, New York, 1976)
Lansing, J. B. et al., Planned Residential Environments (University of Michigan Press, Ann Arbor, 1970)
Lawrence, G. R. P., Randstad Holland (Oxford University Press, London, 1973)
Lineberry, R. L. and Sharkansky, I., Urban Politics and Public Policy (Harper and Row, New York, 1978)
Lijphart, A., The Politics of Accommodation: Pluralism and Democracy (University of California Press, Berkeley and Los Angeles, 1968)
Lopata, H. Z., Occupation Housewife (Oxford University Press, New York, 1971)
Lynch, K. (ed.), Growing Up in Cities: Studies of the Spatial Environment of Adolescence (MIT Press, Cambridge, Mass., and London, 1977)
Maisel, S. J., Housebuilding in Transition (University of California Press, Berkeley, 1953)
Meyer, J. R., Kain, J. F. and Wohl, M., The Urban Transportation Problem (Harvard University Press, Cambridge, Mass., 1972)
Meyerson, M. et al., Housing, People and Cities (McGraw Hill, New York, 1962)
Michelson, W., Man and His Urban Environment (Addison-Wesley, Reading, Mass., 1976)
--- Environmental Choice, Human Behavior and Residential Satisfaction (Oxford University Press, New York, 1977)
Moustakas, C. E., Loneliness and Love (Prentice-Hall, Englewood Cliffs, 1972)
Mumford, L., The City in History (Harcourt, Brace and World, New York, 1961)
Myrdal, G., Economic Theory and Underdeveloped Regions (G. Duckworth and Company, Ltd., London, 1957)
National Advisory Commission on Civil Disorders, Report of the National Advisory Commission on Civil Disorders (Bantam Books, New York, 1968)
Neipris, J., Social Welfare and Social Services in Israel: Policies, Programs and Current Issues (The Hebrew University of Jerusalem, Paul Baerwald School of Social Work, Jerusalem, 1981)
Nycolaas, J., Volkshuisvesting; een bijdrage tot de

geschiedenis van woningbouw en woningbouwbelied (Sun, Nijmegen, 1974)
Oosterbaan, J., Population Dispersal: A National Imperative (Lexington Books, Lexington, Mass., 1980)
Orni, E. and Efrat, E., Geography in Israel (Israel University Press, Jerusalem, 1980)
Pinder, D., The Netherlands (Westview Press, Boulder, 1976)
Popenoe, D., The Suburban Environment: Sweden and the United States (The University of Chicago Press, Chicago, 1977)
Porteous, J. D., Environment and Behavior: Planning and Everyday Urban Life (Addison-Wesley, Reading, Mass., 1977)
Rabier, J., Satisfaction et Insatisfaction Quant Aux Conditions de Vie dans les Pays Membres de la Communauté Européene (Commission of the European Community, Brussels, 1974)
Rea, L. M. and Gupta, D. K., An Economic and Legal Analysis of Rent Control (Institute of Public and Urban Affairs, San Diego State University, San Diego, 1982)
Rein, N., Daughters of Rachel: Women in Israel (Penguin Books, New York 1979)
Rennie, T. A. C. et al., Mental Health in the Metropolis (McGraw Hill, New York, 1962)
Rischin, M., The Promised City: New York's Jews, 1870-1914 Harper and Row, New York, 1970)
Rothblatt, D. N. and Garr, D. J., Comparative Suburban Data (San Jose State University, San Jose, 1983)
--- Planning the Metropolis: The Multiple Advocacy Approach (Praeger, New York, 1982)
--- Garr, D. J., and Sprague, J., The Suburban Environment and Women (Praeger, New York, 1979)
--- (ed.), National Policy for Urban and Regional Development (D. C. Heath, Lexington, Mass., 1974)
Sawers, L. and Tabb, W. K. (eds.), Sunbelt/Snowbelt: Urban Development and Regional Restructuring (Oxford University Press, New York and Oxford, 1984)
Schmandt, H. J. and Bloomberg, W. Jr., Quality of Urban Life (Sage Publications, Beverly Hills, 1969)
Schmitt, P. J. Back to Nature: The Arcadian Myth in Urban America (Oxford University Press, New York, 1969)
Self, P., Planning the Urban Region: A Comparative Study of Policies and Organization (The

University of Alabama Press, Alabama, 1982)
Shevky, E. and Bell, W., Social Area Analysis (Stanford University Press, Stanford, 1955)
Shikun, Housing in Israel (Workers Housing Company Ltd., Tel-Aviv, 1965)
Shimshoni, Daniel, Israeli Democracy: The Middle of the Journey (The Free Press, New York, 1982)
Sicron, M., Immigration to Israel (Falk Project and CBS, Jerusalem, 1957)
Sjoberg, G., The Preindustrial City: Past and Present (Free Press, New York, 1960)
Stanford Environmental Law Society, San Jose: Sprawling City (Stanford Environmental Law Society, Stanford, 1971)
Stein, C., Toward New Towns for America (Reinhold Publishing, New York, 1956)
Stern, G., Israeli Women Speak Out (J. B. Lippincott, Philadelphia and New York, 1979)
Sternlieb, G. and Hughes, J. W. (eds.), Shopping Centers: U.S.A. (Center for Urban Policy Research, New Brunswick, N.J., 1981)
Sundquist, J. L. Dispersing Population: What America Can Learn from Europe (The Brookings Institution, Washington, D.C., 1974)
Szalai, A. and Andrews, F. M. (eds.), The Quality of Life: Comparative Studies (Sage, Beverly Hills and London, 1980)
--- and Petrella, R. (eds.), Cross-National Comparative Survey Research: Theory and Practice (Pergamon Press, New York and London, 1977)
Taylor C. L. and Jodice, D. A., World Handbook of Political and Social Indicators, 3rd edition (Yale University Press, New Haven, 1983)
Te Selle, S. (ed.), The Rediscovery of Ethnicity: Its Implications for Culture and Politics in America (Harper and Row, New York, 1973)
Terleckyi, N. E., Improvements in the Quality of Life: Estimates of Possibilities in the United States (National Planning Asociation, Washington, D.C., 1975)
Thorns, D.C., Suburbia (MacGibbon and Kee, London, 1972)
Van der Schaar, J., Sektorindeling en woningmarkt processen (The Hague: Staatsuitgeverij, 1979)
Van Weesep, Production and Allocation of Housing: The Case of the Netherlands, Geografische en Planologische Notities, no.11 (Free University of Amsterdam, Geografische en Planologische Instituut, Amsterdam, 1982)
Vanhove, N. and Klaassen, L. H., Regional Policy: A

European Approach (Allenheld, Osmun and Company, Montclair, N.J., 1980)
Ward, D., Cities and Immigrants: A Geography of Change in Nineteenth Century America (Oxford University Press, New York, 1971)
Warner, S. B., Streetcar Suburbs, 2nd ed. (Harvard University Press, Cambridge, Mass., 1981)
Weber, A. F., The Growth of Cities in the Nineteenth Century (Cornell University Press, Ithaca, N.Y., 1967)
Weiss, R. E. et al., Loneliness (MIT Press, Cambridge, Mass., 1973)
Werthman, C. et al., Planning and the Purchase Decision: Why People Buy in Planned Communities (University of California Press, Berkeley, 1965)
Wessman, A. E. and Ricks, D.F., Mood and Personality (Holt, Rinehart and Winston, New York, 1966)
Whyte, W. H., The Last Landscape (Doubleday, New York, 1969)
Williams, O. P., Metropolitan Political Analysis (Free Press, New York, 1971)
Wilner, D. M. et al., The Housing Environment and Family Life (Johns Hopkins Press, Baltimore, 1962)
Wood, R. C., Suburbia: Its People and Their Politics (Houghton-Mifflin, Boston, 1958)
--- 1400 Governments (Harvard University Press, Cambridge, Mass., 1961)
Yeates, M., North American Urban Patterns (V. H. Winston and Sons, New York, 1980)
Zehner, R. B., Indicators of the Quality of Life in New Communities (Ballinger Publishing, Cambridge, Mass., 1977)

## ARTICLES AND PERIODICALS

Alden, J. D., 'Metropolitan Planning in Japan', Town Planning Review, vol.55, no.1 (1984), pp.55-74
Allardt, E., 'The Relationship Between Objective and Subjective Indicators in the Light of a Comparative Study', Comparative Studies in Sociology, vol.1, no.2 (1978), pp.203-15
Allen, J. B., 'The Next Seven Years: A Real Estate Perspective', Grubb and Ellis Investor Outlook, vol.4, no.2 (1984), pp.1-3
Alonso, W., 'A Theory of the Urban Land Market', in M. Edel and J. Rothenberg (eds.) Readings in Urban Economics (MacMillan, New York, 1972), pp.104-11

--- 'Location Theory', in J. Friedmann and W. Alonso (eds.) Regional Policy: Readings in Theory and Applications (MIT Press, Cambridge, Mass., 1975) pp.35-63

Andrews, F. M., 'Comparative Studies of Life Quality: Comments on the Current State of the Art and Some Issues for Future Research', in A. Szalai and F. M. Andrews (eds.) The Quality of Life: Comparative Studies (Sage, Beverly Hills and London, 1980), pp.273-85

Baldassare, M. and Fischer, C.S., 'Suburban Life: Powerlessness and Need for Affiliation', Urban Affairs Quarterly, vol.10, no.3 (1975), pp.314-26

Bell, W. and Boat, M., 'Urban Neighborhoods and Informal Social Relations', American Journal of Sociology, vol.62, no.3 (1957), pp.391-8

Belser, K., 'The Making of Slurban America', Cry California, vol.5, no.4 (1970), pp.1-21

Berdichevsky, N., 'The Persistence of the Yemeni Quarter in an Israeli Town', in E. Krausz (ed.) Studies of Israeli Society: Volume 1 (Transaction Books, London, 1980), pp.73-95

Berry, B. J. L., 'The Counterurbanization Process: Urban America Since 1970', in B. J. L. Berry (ed.), Urbanization and Counterurbanization (Sage, Beverly Hills and London, 1976), pp.17-30

Borchert, J. G., 'The Dutch Settlement System', in H. Van der Haegen (ed.) Western European Settlement Systems (Instituut voor Sociale en Economische Geografie Katholike Universiteit te Lueven, Lueven, 1982), pp.207-50

Borukhov, E., Ginsberg, Y. and Werczberger, E., 'Housing Prices and Housing Preferences in Israel', Urban Studies, vol.15, no.2 (1978), pp.187-200

Bourne, L. S. and Logan, M. I., 'Changing Urbanization Patterns at the Margin: The Examples of Australia and Canada', in B. J. L. Berry (ed.) Suburbanization and Counterurbanization (Sage, Beverly Hills and London, 1976), pp.111-44

Boyd, M. et al., 'Status Attainment of Immigrant and Immigrant Origin Categories in the United States, Canada, and Israel', Comparative Social Research, vol.3, no.2 (1980), pp.199-227

Brand, W., 'The Legacy of Empire', in R. T. Griffiths (ed.), The Economy and Politics of the Netherlands Since 1945 (Martinus Nijhoff, The Hague, 1980), pp.251-75

Buder, S., 'The Future of the American Suburbs', in P. C. Dolce (ed.), Suburbia: The American Dream and Dilemma (Anchor Books, Garden City, N.Y., 1976), pp.193-216

Bylinsky, G., 'California's Great Breeding Ground for Industry', Fortune, vol.89, no.6 (1974), pp.129-35, 216-24

Carmon, N. and Mannheim, B., 'Housing as a Tool of Social Policy', Social Forces, vol.58, no.2 (1979), pp.336-51

Checkoway, B., 'Large Builders, Federal Housing Programmes, and Postwar Suburbanization', International Journal of Urban and Regional Research, vol.4, no.1 (1980), pp.21-44

Cherry, G. E., 'Britain and the Metropolis: Urban Change and Planning in Perspective', Town Planning Review, vol.2, no.55 (1984), pp.5-33

Clark, W. A. V. and Everaers, P. C. J., 'Public Policy and Residential Mobility in Dutch Cities', Tijdschrift voor Economische en Sociale Geografie, vol.72, no.6 (1981), pp.322-33

Clout, H. D., 'Population and Urban Growth', in H. D. Clout (ed.), Regional Development in Western Europe, (John Wiley, New York, 1981), pp.35-59

Comay, Y. and Kirchenbaum, A., 'The Israeli New Town: An Experiment of Population Redistribution', Economic Development and Cultural Change vol.22, no.1 (1974), pp.124-34

Dahmann, D. C., 'Subjective Assessment of Neighborhood Quality by Size', Urban Studies, vol.20, no.1 (1983), pp.31-45

Darin-Drabkin, H., 'Economic and Social Aspects of Israeli Housing', in H. Darin-Drabkin (ed.) Public Housing in Israel; Surveys and Evaluations of Activities in Israel's First Decade (1948-1958) (Gadish Books, Tel-Aviv, 1959), pp.15-92

Davidoff, P. and Brooks, M., 'Zoning Out the Poor', in P. C. Dolce (ed.), Suburbia: The American Dream and Dilemma (Anchor Press, New York, 1976), pp.135-66

de Wolff, P. and Driehuis, W., 'A Description of Post War Economic Developments and Economic Policy in the Netherlands', in R. T. Griffiths (ed.) The Economy and Politics of the Netherlands Since 1945 (Martinus Nijhoff, The Hague, 1980), pp.13-60

Downes, B. T., 'Suburban Differentiation and Municipal Policy Choices', in T. N. Clark

(ed.), Community Structure and Decision-Making (Chandler Publishing, San Francisco, 1968), pp.243-67
Erickson, J. A., 'An Analysis of the Journey to Work for Women', Social Problems, vol.24, no.4 (1977), pp.428-35
Fava, S. F., 'Beyond Suburbia', Annals of the American Academy of Political and Social Science, vol. 422 (1975), pp.10-24
Feldman, A. and Tilly, C., 'The Interaction of Social and Physical Space', American Sociological Review, vol.25, no. 6 (1960), pp.877-84
Fielding, A. J., 'Counterurbanization in Western Europe', Progress in Planning, vol.17, part 1 (1982), pp.1-45
Fischer, C. S. and Jackson, R. M., 'Suburbs, Networks and Attitudes', in B. Schwartz (ed.) The Changing Face of the Suburbs (The University of Chicago Press, Chicago, 1976), pp.279-309
--- 'Toward a Subcultural Theory of Urbanism', American Journal of Sociology, vol.80, no.6 (1975), pp.1319-41
Gallup, G. H., 'Human Needs and Satisfactions: A Global Survey', Public Opinion Quarterly, vol.40, no.4 (1976), pp.459-67
Garling, T., Book, A. and Lindsberg, E., 'Cognitive Mapping of Large Scale Environments: The Interrelationship of Action Plans, Acquisition, and Orientation', Environment and Behavior, vol.16, no.1 (1984), pp.3-34
Gay, F. J., 'Benelux', in H. D. Clout (ed.), Regional Development in Western Europe (John Wiley, New York, 1981), pp.179-209
Glazer, N., 'Slum Dwellings Do Not Make A Slum', New York Times Magazine (November 21, 1965), p.55
Glikson, A., 'Some Problems in Housing in Israel's New Towns and Suburbs', in H. Darin-Drabkin (ed.), Public Housing in Israel: Surveys and Evaluations of Activities in Israel's First Decade (1948-1958) (Gadish Books, Tel-Aviv, 1959), pp.93-102
--- 'Urban Design in New Towns and Neighborhoods', Ministry of Housing Quarterly (December 1967), pp.45-51
Goldfield, D. R., 'National Urban Policy in Sweden', Journal of the American Planning Association, vol.48, no.1 (1982), pp.24-38
Gonen, A., 'The Suburban Mosaic in Israel', in D. H. K. Amiram and Y. Ben-Arieh (eds.)

Geography in Israel (The Israel National Committee, Jerusalem, 1976), pp.163-86
Gotthiel, F. M. 'On the Economic Development of the Arab Region in Israel', in M. Curtis and M. S. Chertoff (eds.), Israel: Social Structure and Change (Transaction Books, New Brunswick, N.J., 1973), pp.237-48
Gove, W. R. and Tudor, J. F., 'Adult Sex Roles and Mental Illness', in J. Huber (ed.) Changing Women in a Changing Society (University of Chicago Press, Chicago, 1973), pp.50-73
Graizer, I., 'Spatial Patterns and Residential Densities in Israeli "Moshavot" in Process of Urbanization', Geo Journal, vol.2, no.6 (1978), pp.533-7
Greer, S., 'The Family in Suburbia', in L. H. Masotti and J. K. Hadden (eds.) The Urbanization of the Suburbs (Sage, Beverly Hills and London, 1973), pp.149-70
Grimshaw, A. D., 'Comparative Sociology: In What Ways Different from Other Sociologies?' in M. Armer and A. D. Grimshaw (eds.) Comparative Social Research: Methodological Problems and Strategies (John Wiley, New York, 1973) pp.3-48
Guest, A. M. and Lee, B. A., 'The Social Organization of Local Areas', Urban Affairs Quarterly, vol.19, no.2 (1983), pp.217-40
Gulick, L., 'Metropolitan Organization', Annals of the American Academy of Political and Social Science, vol.314 (1957), pp.57-65
Hamnett, S., 'The Netherlands: Planning and the Politics of Accommodation', in D. H. MacKay (ed.), Planning and Politics in Western Europe (St. Martin's Press, New York, 1982), pp.111-43
Hankiss, E., 'Structural Variables in Cross-Cultural Research on the Quality of Life', in A. Szalai and F. M. Andrews (eds.), The Quality of Life: Comparative Studies (Sage, Beverly Hills and London, 1980), pp.46-56
Harris, C. D., 'The Urban and Industrial Transformation of Japan', The Geographic Review, vol.72, no.1 (1982), pp.50-89
Hartman, H. and Hartman, M., 'The Effect of Immigration on Women's Roles in Various Countries', International Journal of Sociology and Social Policy, vol.3, no.3 (1983), pp.86-103
Hasson, S., 'The Emergence of an Urban Social Movement in Israeli Society - An Integrated Approach', International Journal of Urban and Regional Research, vol.7, no.2 (1983),

157-74

Havens, E. M. 'Women, Work and Wedlock: A Note on Female Marital Patterns in the United States', in J. Huber (ed.), Changing Women in a Changing Society (The University of Chicago Press, Chicago, 1973), pp.213-9

Izraeli, D. N., 'Sex Structure of Occupations: The Israeli Experience', Sociology of Work and Occupations vol.6, no.4 (1979), pp.404-29

Jackson, K. T., 'Race, Ethnicity and Real Estate Appraisal: The Home Owners Loan Corporation and the Federal Housing Administration', Journal of Urban History, vol.6, no.4, (August 1980), pp.419-52

Johnson, J. A., 'Geographical Processes at the Edge of the City', in J. H. Johnson (ed.) Suburban Growth (John Wiley and Sons, New York, 1974), pp.1-16

Kain, J. F., 'The Distribution and Movement of Jobs and Industry', in J. O. Wilson (ed.), The Metropolitan Enigma (United States Chamber of Commerce, Washington, D.C., 1967), pp.1-31

--- and Persky, J. J., 'Alternatives to the Gilded Ghetto', The Public Interest, no.12 (Winter 1969), pp.74-87

--- 'The Journey to Work as a Determinant of Residential Location', in A. N. Page and W. R. Siegfried (eds.) Urban Analysis (Scott, Foresman, Glenview, 1970), pp.207-26

Kaniss, P. and Robins, B., 'The Transportation Needs of Women', in K. Hapgood and J. Getzels (eds.) Planning, Women and Change (American Society of Planning Officials, Chicago, 1974), pp.63-70

Kantrowitz, N., 'Ethnic and Racial Segregation in the New York Metropolis, 1960', American Journal of Sociology, vol.74, no.6 (1969), pp.685-95

Khorev, B. S. and Moiseenko, V. M., 'Urbanization and Redistribution of the Population of the U.S.S.R.', in S. Goldstein and D. S. Sly (eds.), Patterns of Urbanization: Comparative Country Studies, Volume 2 (International Union for the Scientific Study of Population, Dolhain, Belgium, 1977), pp.643-720

Klaff, V., 'Residence and Integration in Israel: A Mosaic of Segregated Groups', in E. Krauz (ed.), Studies of Israeli Society, Volume I (Transaction Books, London, 1980), pp.53-71

Klein, P. W., 'The Foundations of Dutch Prosperity', in R. T. Griffiths (ed.), The Economy and Policy of the Netherlands Since 1945 (Martinus

Nijhoff, The Hague, 1980), pp.1-12
Kombrink, H., 'Huidig doorstromingsbelied is te weinig op behoeften afgestemd', Bouw, no.25 (1978), pp.46-8
Kuroda, T., 'The Impact of Internal Migration on the Tokyo Metropolitan Region', in J. W. White (ed.), The Urban Impact of Internal Migration (Institute for Research in Social Science, University of North Carolina, Chapel Hill, 1979), pp.33-52
Lappk, D. and Van Hoogtraten, P., 'Remarks on the Spatial Structure of Capitalist Development: The Case of the Netherlands', in J. Carney, R. Hudson and J. Lewis (eds.), Regions in Crisis: New Perspectives in European Regional Theory (St. Martin's Press, New York, 1980), pp.117-71
Leven, C. L. 'Regional Variations in Metropolitan Growth and Development', in V. L. Arnold (ed.), Alternatives to Confrontation: A National Policy Towards Regional Change (Lexington Books, Lexington, Mass., 1980), pp.329-43
Lichfield, N., 'The Israeli Physical Planning System: Some Needed Changes', The Israel Annual of Public Adminstration, vol.15 (1976), pp.35-58
Logan, J. R. and Schneider, M., 'Governmental Organization and City/Suburb Income Inequality, 1960-1970', Urban Affairs Quarterly, vol.17, no.3 (1982), pp.303-18
London, B., Bradley, D. S. and Hudson, J.R., 'The Revitalization of Inner-City Neighborhoods', Urban Affairs Quarterly, vol.15, no.4 (1980), pp.373-80
Lucy, W. H., 'Metropolitan Dynamics: A Cross-National Framework for Analyzing Public Policy Effects in Metropolitan Areas', Urban Affairs Quarterly, vol.11, no. 2 (December 1975), pp.155-85
Maas, M. W. A., 'Condomiumium Conversion in Pre-War Neighborhoods: An Urban Transformation Process in Dutch Cities', Tijdschrift voor Economische en Sociale Geografie, vol.76 (1984), pp.36-45
Maisel, S. and Winnick, L., 'Family Housing Expenditures: Illusive Laws and Intrusive Variances', in W. L. C. Wheaton, et al. (eds.), Urban Housing (Free Press, New York, 1966), pp.139-53
Marshall, H., 'Suburban Life Styles: A Contribution to the Debate', in L. Masotti and J. K. Hadden (eds.) The Urbanization of the Suburbs (Sage, Beverly Hills and London, 1973), pp.123-48

Masotti, L. H. and Bowen, D., 'Communities and Budgets: The Sociology of Municipal Expenditures', Urban Affairs Quarterly, vol.1, no. 1 (1965), pp.38-58

--- 'Suburbia Reconsidered: Myth and Counter-Myth', in L. H. Masotti and J. K. Hadden (eds.) The Urbanization of the Suburbs (Sage, Beverly Hills and London, 1973), pp.15-22

McKay, D. H. 'Planning in the Mixed Economy: Problems and Prospects', in D. H. McKay (ed.) Planning and Politics in Western Europe (St. Martin's Press, New York, 1982), pp.170-85

Meyer, J. R., 'Urban Transportation', in J. Q. Wilson (ed.) The Metropolitan Enigma (United States Chamber of Commerce, Washington, D.C., 1967), pp.34-75

Moore, G. T., 'Knowing about Environmental Knowing: The Current State of Theory and Research on Environmental Cognition', Environment and Behavior, vol.11, no.2 (1979), pp.33-70

Novak, M., 'How American Are You If Your Grandparents Came from Serbia in 1888?' in S. Te Selle (ed.) The Rediscovery of Ethnicity: Its Implications for Culture and Politics in America (Harper and Row, New York, 1973), pp.1-20

Nowak, S., 'The Strategy of Cross-National Survey Research for the Development of Social Theory', in A. Szalai and R. Petrella (eds.), Cross National Comparative Survey Research: Theory and Practice (Pergamon Press, New York and London, 1977), pp.3-48

Osmund, H., 'Some Psychiatric Aspects of Design', in L. B. Holland (ed.), Who Designs America? (Anchor Books, Garden City, 1966), pp.281-318

Palen, J. J., 'The Urban Nexus: Toward the Year 2000', in A. A. Hawley (ed.), Societal Growth (Free Press, New York, 1979), pp.141-56

Perloff, H. S., 'A Framework for Dealing with the Urban Environment: An Introductory Statement', in The Quality of the Urban Environment (Johns Hopkins Press, Baltimore, 1969), pp.3-31

Pfeil, E., 'The Pattern of Neighboring Relations in Dortmund-Norstadt', in R. E. Pahl (ed.) Readings in Urban Sociology (Pergamon Press, London, 1968), pp.136-58

Plant, S. E., 'The Economics of Population Dispersal', Urban Studies, vol.20, no.3 (August 1983), pp.353-7

Popenoe, D., 'Urban Form in Advanced Societies: A Cross-National Enquiry', in C. Ungerson and

V. Karn (eds.), The Consumer Experience of Housing: Cross National Perspectives (Gower, Aldershot, Hants., 1980), pp.1-20
Pray, M. M., 'Planning and Women in the Suburban Setting', in K. Hapgood and J. Getzels (eds.) Planning, Women and Change (American Society of Planning Officials, 1974), pp.50-60
Priemus, H., 'Rent Control and Housing Tenure', Planning and Administration, vol.9, no.2 (1982), pp.29-46
Rabinovitz, F. and Lamare, J., 'After Suburbia, What?' in W. Z. Hirsch (ed.), Los Angeles: Viability and Prospects for Metropolitan Leadership (Praeger, New York, 1971), pp.169-206
Raven, J., 'Sociological Evidence on Housing (2: The Home Environment)', The Architectural Review, vol.142, no.1 (1967), pp.236-7
Romann, M., 'Jews and Arabs in Jerusalem', The Jerusalem Quarterly, no.19 (Spring 1981), pp.23-46
Rosenberg, B., 'Women's Place in Israel: Where They Are, Where They Should Be', Dissent, vol.24, no.4 (1977), pp.408-17
Rothblatt, D. N. 'Multiple Advocacy: An Approach to Metropolitan Planning', Journal of the American Institute of Planners, vol.44, no.2 (1978), pp.193-9
--- 'Improving the Design of Urban Housing', in V. Kouskoulas (ed.), Urban Housing (National Sciences Foundation, Detroit, 1973), pp.149-54
--- 'Housing and Human Needs', Town Planning Review, vol.42, no.2 (1971), pp.130-44
Rushing, W., 'Two Patterns in the Relationship Between Social Class and Mental Hospitalization', American Sociological Review, vol.34, no.4 (1966), pp.533-41
Saegert, S., 'Masculine Cities and Feminized Suburbs: Polarized Ideas, Contradictory Realities', Signs, vol.5, Spring supplement (1980), pp.96-111
Sanoff, H., 'Neighborhood Satisfaction: A Study of User Assessments of Low Income Residential Environment', in O. Ural (ed), Proceedings of the Second International Symposium on Lower Cost Housing Problems (University of Missouri-Rolla, St. Louis, 1972), pp.119-24
Saxenian, A., 'The Urban Contradictions of Silicon Valley: Regional Growth and the Restructuring of the Semiconductor Industry', in L. Sawers and W. K. Tabb (eds.) Sunbelt/Snowbelt: Urban

Development and Regional Restructuring (Oxford University Press, New York and Oxford, 1984), pp.163-97

Schmandt, H. J. and Stephens, G., 'Measuring Municipal Output', National Tax Journal, vol.13, no.4 (1960), pp.369-75

Schulman, N., 'Mutual Aid and Neighboring Patterns: The Lower Town Study', Anthropoligica, vol.9, no.1 (1967), pp.51-60

Schwirian, K. P., 'Some Analytical Problems in the Comparative Test of Ecological Theories', in M. Armer and A. D. Grimshaw (eds.) Comparative Social Research: Methodological Problems and Strategies (John Wiley, New York, 1973), pp.347-72

Semyonov, M. and Kraus, V., 'Gender, Ethnicity and Income Inequality', International Journal of Sociology, vol.24, no.3 (1983), pp.258-72

Shachar, A., 'New Towns in a National Settlement Policy', Town and Country Planning, vol.44, no.2 (1976), pp.83-7

Shaham, I., 'Public Housing in Israel', in J. S. Fuerst (ed.) Public Housing in Europe and America (John Wiley and Sons, New York, 1974), pp.52-66

Short, J. R., 'Urban Policy and British Cities', Journal of the American Planning Association, vol.48, no.1 (1982), pp.39-52

Smooha, S. and Peres, Y., 'The Dynamics of Ethnic Inequalities: The Case of Israel', in E. Krauz (ed.), Studies of Israeli Society, Volume I (Transaction Books, London, 1980), pp.165-81

Soen, D., 'Israel's Population Dispersal Plans and Their Implication, 1949-74: Failure or Success?' Geo Journal, vol.1, no.5 (1977), pp.378-81

Starr, R. E., 'Infill Development - Opportunity or Mirage', Urban Land, vol.39 (1980), pp.3-5

Stipak, B., 'Citizen Satisfaction with Urban Services: Potential Misuse as a Performance Indicator', Public Administration Review, vol.39, no.1 (1979), pp.46-52

Tarr, J. A., 'From City to Suburb: The "Moral" Influence of Transportation Technology', in A. B. Callow (ed.) American Urban History (Oxford University Press, New York, 1973), pp.201-23

Tiebout, C. M., 'A Pure Theory of Local Expenditures', in M. Edel and J. Rothenberg (eds.) Readings in Urban Economics (MacMillan, New York, 1972), pp.513-23

Tomeh, A. K., 'Informal Group Participation and Residential Patterns', American Journal of Sociology, vol.70, no.1 (1964), pp.28-35

Van de Bergh, F., 'De koers is om maar het schip zal stranden', Bouw, no.25 (1978), pp.17-21

Van de Kaa, D. J., 'Population Prospects and Population Policy in the Netherlands', Netherlands Journal of Sociology, vol.17 (1981), pp.73-91

Van der Knaap, G. A., 'Sectoral and Regional Imbalances in the Dutch Economy', in R. T. Griffiths (ed.), The Economy and Politics of the Netherlands Since 1945 (Martinus Nijhoff, The Hague, 1980), pp.115-34

Van Gelder, L., 'Strains of "Oh Promise Me" Emanating from the White House', San Jose News (February 8, 1979)

Van Weesep, J., 'Intervention in the Netherlands: Urban Housing Policy and Market Response', Urban Affairs Quarterly, vol.19, no.3 (March 1984), pp.329-53

Vining, D. R. Jr. and Kontuly, T., 'Population Dispersal for Major Metropolitan Regions: An International Comparison', International Regional Science Review, vol.3, no.1 (1978), pp.49-73

Ward, Z. A., 'A Policy Antimony: Public Attitudes Versus Urban Conditions in Western Europe', in M. C. Romanos, Western European Cities in Crisis (Lexington Books, Lexington, Mass., 1979), pp.47-66

Weissman, M. M. and Paykel, E. S., 'Moving and Depression in Women', in R. S. Weiss (ed.), Loneliness (MIT Press, Cambridge, Mass., 1973), pp.154-64

Wiatr, J., 'The Role of Theory in the Process of Cross-National Survey Research', in A. Szalai and R. Petrella (eds.), Cross-National Comparative Survey Research: Theory and Practice (Pergamon Press, New York, 1977), pp.347-72

Wolforth, J., 'The Journey to Work', in L. S. Bourne (ed.), Internal Structure of the City (Oxford University Press, New York, 1971), pp.240-7

Wright, J. D., 'Are Working Women Really More Satisfied?' Journal of Marriage and the Family, vol.40, no.2 (1978), pp.301-14

Zelan J., 'Does Suburbia Make a Difference?' in S. F. Fava (ed.) Urbanism in World Perspective (Crowell, New York, 1968), pp.401-8

Zimmer, B. G., 'The Urban Centrifugal Drift', in

A. H. Hawley and V. P. Rock (eds.), Metropolitan America in Contemporary Perspective (Halstead Press/John Wiley, New York and London, 1975), pp.23-92

PUBLIC DOCUMENTS

Association of Bay Area Governments, Projections 79 (Association of Bay Area Governments, Berkeley, 1979)

Bogue, D. J., Population Growth in Standard Metropolitan Areas 1900-1950 (Housing and Home Finance Agency, Washington, D.C., 1953)

Great Britain, Department of Statistics, Statistical Abstract of Palestine: 1944-45 (Government Printer, London, 1946)

Israel Central Bureau of Statistics, Statistical Abstract of Israel: 1978 (Silvan Press, Jerusalem, 1978)

Israel Central Bureau of Statistics, Statistical Abstract of Israel 1979 (Kiryat Arba Press, Hebron, 1980)

Israel Ministry of Housing, Population and Building in Israel: 1948-73 (Jerusalem, 1975)

Long, J. F., Population Decentralization in the United States (U.S. Department of Commerce, Bureau of the Census, Washington, D.C., 1981)

--- Population Deconcentration in the United States (U.S. Department of Commerce, Bureau of the Census, Washington, D.C., 1981)

National Advisory Commission on Civil Disorders, Report of the National Advisory Commission on Civil Disorders (Bantam Books, New York, 1968)

Netherlands Central Bureau of Statistics, Statistical Yearbook of the Netherlands: 1980 (Staatsuitgeverij, The Hague, 1981)

Netherlands Minister of Economic Affairs,'Memorandum on the Industrialization of the Netherlands', (September 1949) cited in J. G. Abert Economic Policy and Planning in the Netherlands, 1950-1965 (Yale University Press, New Haven, 1969)

Netherlands Ministry of Housing and Physical Planning, Summary of the Orientation Report on Physical Planning (Staatsuitgeverij, The Hague, 1974)

--- Summary of the Report on Urbanization in the Netherlands (Staatsuitgeverij, The Hague, 1976)

--- The Relationship Between Physical Planning, Policy and Economic Policy (Staatsuitgeverij, The Hague, 1977)

--- Housing Production in the Netherlands (Staatsuitgeverij, The Hague, 1978)
Netherlands Social and Cultural Planning Office, Social and Cultural Report: 1978 (Rijswijk, 1978)
Santa Clara County Housing Task Force, Housing: A Call for Action (Santa Clara County Planning Department, San Jose, 1977)
Santa Clara County Planning Department, Advanced Final Count of 1980 (San Jose, April 1981)
Santa Clara County Planning Department, 1975 Countywide Census (San Jose, 1976)
Santa Clara County Planning Department, Housing Characteristics, Cities, Santa Clara County, 1970 (San Jose, 1971)
United Nations, Patterns of Urban and Rural Population Growth (United Nations, New York, 1980)
U.S. Department of Commerce, Bureau of the Census, Census of Population 1950 (Washington, D.C., 1950)
U.S. Department of Commerce, Bureau of the Census, Census of Population 1960 (Washington, D.C., 1960)
U.S. Department of Commerce, Bureau of the Census, Census of Population 1970 (Washington, D.C., 1970)
U.S. Department of Commerce, Bureau of the Census, Household Income in 1972 and Selected Social and Economic Characteristics of Households (Washington, D.C., 1972)
U.S. Department of Commerce, Bureau of the Census, Population and Land Areas of Urbanized Areas for the United States and Puerto Rico (Washington, D.C., 1984)
U.S. Department of Commerce, Bureau of the Census, Preliminary 1980 Census (Washington, D.C., March 1981)
U.S. Geological Survey, "San Francisco Bay Region Map" (1970)

UNPUBLISHED MATERIAL

Carney, J. M., 'How to Evaluate the Impacts of the Combined General Plans of the Cities of Santa Clara County, California', Master's Planning Report, San Jose State University, 1978
de Jonge, D., 'Some Notes on Sociological Research in the Field of Housing', mimeograph, Delft University of Technology, 1967
Freitas, M. 'Women in Suburbia', Master's Planning

Report, San Jose State University, 1974
Gabriel, S. and Maoz, I., 'Cyclical Fluctuations in the Israel Housing Market', Center for Real Estate and Urban Economics, Graduate School of Business, University of California, Berkeley (June 1983)
Gradus, Y., 'The Role of Politics in Regional Inequity in Israel', paper delivered at the International Seminar on Contemporary Problems in Political Geography, Haifa, Israel, May 1982
Gottdiener, 'The New Form of Settlement Space: Conceptualizing Decentralization', paper delivered at the Association of Collegiate Schools of Planning Conference, San Francisco, October 1983
Inkeles, A. and Diamond, L., 'Personal Qualities as a Reflection of Level of National Development', paper delivered at the Ninth World Congress of the International Sociological Association, Uppsala, Sweden, August 1978
Jones, L. M., 'The Labor Force Participation of Married Women', Master's Thesis, University of California, Berkeley, 1974
Lev-Ari, A., 'Spatial Behavior and Overt Residential Preference', unpublished Master's Thesis, Hebrew University of Jerusalem, 1978
Michelson, W., 'Environmental Change', Centre for Urban and Community Studies, Research Paper No.60 (October 1973)
Rothblatt, D. N. and Garr, D. J., 'Suburbia: An International Perspective', paper delivered at the Association of Collegiate Schools of Planning Conference, New York, October 1984

## ABOUT THE AUTHORS

Donald N. Rothblatt chairs the Urban and Regional Planning Department at San Jose State University and is a Research Associate at the Institute of Government Studies, University of California, Berkeley. A former president of the Association of Collegiate Schools of Planning, his works include *Human Needs and Public Housing*, *Thailand's Northeast*, *Regional Planning: The Appalachian Experience*, *Allocation of Resources for Regional Planning*, *National Policy for Urban and Regional Development*, *Planning the Metropolis: The Multiple Advocacy Approach*, and (as co-author) *The Suburban Environment and Women*. He has studied and practised planning in the United States, Europe, and the Middle East, and holds the Ph.D. in city and regional planning from Harvard University, where he was on the planning faculty.

Daniel J. Garr is Professor of Urban and Regional Planning at San Jose State University. He has practised planning in both rural and urban regions and has published widely in the area of urban history and social policy, including the co-authorship of *The Suburban Environment and Women*, and *Spanish City Planning in North America*. A winner of the Herbert E. Bolton Award of the Western History Association, he was a Senior Fulbright Research Scholar at the Institute for Town Planning Research, Delft University of Technology in the Netherlands. Dr. Garr holds graduate planning degrees from the University of California, Berkeley, and Cornell University.

SUBJECT INDEX

## NAME INDEX

Ingram Content Group UK Ltd.
Milton Keynes UK
UKHW020053290623
424233UK00007B/30